Albert Einstein's Vision

Albert Einstein's Vision

Remarkable Discoveries

That Shaped

Modern Science

Barry Parker

Prometheus Books

59 John Glenn Drive
Amherst, New York 14228-2197

Published 2004 by Prometheus Books

Inquiries should be addressed to
Prometheus Books
59 John Glenn Drive
Amherst, New York 14228–2197
VOICE: 716–691–0133, ext. 207
FAX: 716–564–2711
WWW.PROMETHEUSBOOKS.COM

08 07 06 05 04   5 4 3 2 1

Library of Congress Cataloging-in-Publication Data

Parker, Barry.
    Albert Einstein's vision : remarkable discoveries that shaped modern science / Barry Parker.
        p. cm.
    Includes bibliographical references and index.
    ISBN 1–59102–186–3 (hardcover : alk. paper)
    1. Einstein, Albert, 1879–1955—Influence. 2. Relativity (Physics) 3. Quantum theory. 4. Physicists—Intellectual life. I. Title.
QC16.E5P365 2005
530.11—dc22

2004003990

Printed in Canada on acid-free paper

Contents

Preface

I have written several books on Einstein, but as I sit here thinking about this book, I am still amazed at the breadth of his contributions to science. It is incredible that one person could affect science so profoundly. His insights were, indeed, miraculous. This book is the third in a trilogy. The first was *Einstein's Brainchild: Relativity Made Relatively Easy*, which was directed mostly at Einstein's two great theories: his special and general theories of relativity. The second was *Einstein: The Passions of a Scientist*. It was directed at the private life of Einstein—in particular, his strong passions for research, music, women, and peace. The present book is directed mainly at Einstein's "other contributions," in other words, those other than special and general relativity. In particular, I discuss how Einstein's other ideas have affected and shaped modern science. Among these contributions are his work on black holes, gravitational waves, cosmic lenses, statistics, and his important contributions to quantum theory and unified field theory. Unified field theory, or more generally, the search for a theory of everything, is now an important endeavor in physics. Einstein was the first to take up the challenge. In this book I outline the important advances that have been made in this area within the past few years.

It is not possible to write about science without occasionally using scientific terms. As much as possible, I have tried to explain them as they cropped up in the text, but for the benefit of those with little scientific background, I have included a glossary at the back of the book.

Very large and very small numbers are also a problem in a popular science book, and I have occasionally used scientific notation in an effort to get around them. In this notation a number such as 100,000,000 is written as 10^8 (the exponent gives the number of zeros). Similarly, small numbers such as 1/100,000,000 are written as 10^{-8}. I have also used the Kelvin, or absolute, temperature scale, in places. On this scale, 0K is the lowest temperature in the universe; in the Fahrenheit scale it is −459°F.

The line drawings and sketches were done by Lori Scoffield-Beer. I would like to thank her for an excellent job. I would also like to thank my editor, Linda Greenspan Regan, and the staff of Prometheus Books for the help in bringing this book to its final form. Finally, I would like to thank my wife for her support while it was being written. To see descriptions of my other books visit my Web site, www.BarryParkerbooks.com.

Introduction

E instein and his friend Michele Besso were walking home from work late in the afternoon. "I think I'm getting close to the real meaning of space and time, but I still have a problem," Einstein said to Besso. He stopped as he explained the problem to Besso in more detail. Besso nodded as he listened. Within minutes the two men parted, and Einstein continued home deep in thought. The next morning Einstein greeted Besso. "Thanks, I have solved my problem," he said.[1]

This was one of the first of Einstein's amazing breakthroughs, and it would soon lead to the special theory of relativity. Indeed, four more breakthroughs would occur in the same year. And for the next twenty-five years Einstein would continue to make mind-boggling discoveries, his greatest coming in 1915 with the formulation of the general theory of relativity. It was almost as if he had a direct pipeline to what some call the "mind of God."

It is remarkable that Einstein, almost single-handedly, laid the foundation of the two major scientific revolutions that occurred in the twentieth century. And these were only two of his numerous contributions,

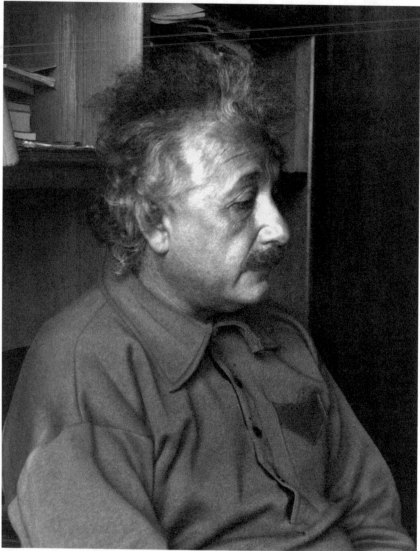

Fig. 1: Einstein in his study. (Photo courtesy of the Lotte Jacobi Archives.)

with some of them most assuredly worthy of the Nobel Prize. (He was, in fact, nominated for the Nobel Prize for several of them.) It's not an exaggeration to say that his discoveries have shaped almost all areas of modern physics. Many new branches of physics have, in fact, arisen as a result of his work. Prominent among them are black hole physics, gravitational wave research, cosmic lensing, and the study of the creation and evolution of our universe.

Einstein's discoveries are still making headlines. Shortly after he formulated his general theory of relativity in 1915, Einstein applied his theory to the universe. He was surprised when his equations told him that the universe was unstable—it was either collapsing or expanding. To overcome this, he introduced a term he called the "cosmological constant." He was reluctant to use the term because he felt it destroyed the beauty of his equations, but when he did, it gave him a static model of the universe. In 1932, however, after it was discovered that the universe was expanding, he rejected the constant, calling it the biggest blunder of his life. Interestingly, the term is now back in vogue. Astronomers have discovered that the expansion of the universe is not slowing down as predicted, and it appears that Einstein's cosmological constant is needed to explain it.

Another of Einstein's discoveries made headlines in the late 1990s. In 1932 he predicted an exotic new form of matter. In 1995 scientists created this exotic new matter, now called a "condensate," in the lab for the first time, and it continues to generate enormous interest.

After formulating his special theory of relativity in 1905, Einstein realized it was not complete. It gave us a new understanding of space, time, and motion, but it was restricted to straight-line, uniform motion. He wanted to generalize it to include all types of motion, and it was here that he made his greatest contribution to science. General relativity, as his new theory was called, gave us a new explanation of gravity in terms of something that seemed outrageous to many. He suggested that gravity was curved space, a concept we could not see and could deal with only through mathematics. Nevertheless, it was an amazingly accurate and beautiful theory that explained many things that Isaac Newton's theory of gravity could not explain. In addition to elucidating the motions within the solar system and beyond, it helped us understand the creation and evolution of the universe.[2]

One of the more bizarre predictions to come out of general relativity was what we now call a black hole. Early on, Einstein and others noted that, according to general relativity, a spherical object such as a star had a "singularity" at a finite distance from the center.[3] This meant that if all the mass of the star was somehow squeezed down to this radius, strange things would happen. One of them was that no light would be emitted from the star, and as a result we would not be able to see it. (We might, however, be able to detect it indirectly since it would block the light from background objects and would appear in a field of stars as a black sphere.) This was only one of its many strange properties. As we will see, many of them are so weird that they are hard to fathom.

Singularities were not new to science. They are regions where, according to the equations that describe them, a theory breaks down and does not give us a finite answer. Singularities occur in both Newton's gravitational theory and James Clerk Maxwell's electromagnetic theory at the center of an object; in other words, at zero radius. This had been known for years and was accepted as a slight, but not serious, flaw in both theories. But to have a singularity at a finite radius was something new. It was bizarre, to say the least. Einstein was disturbed when he realized such a singularity existed in general relativity. He was sure it was a serious flaw in the theory, and he struggled for years to eliminate it. Then in 1939 J. Robert Oppenheimer showed that the singularity had physical relevance when applied to a collapsing star. According to Oppenheimer's calculations, a strange object, later called a black hole, would be created.[4] Einstein refused to believe it. But after World War II interest in these strange new objects blossomed, and a new branch of physics, namely, black hole physics, was formed. Scientists were fascinated with the strange, new objects, and the new science was given a boost when several good candidates were found in space.

It is perhaps ironic that in his struggle to rid the theory of singularities, Einstein made a major discovery about the properties of space and time. Space could become so warped that it twisted back on itself, creating a "bridge" through space and time. We now refer to these bridges as "wormholes" in space. They are known to be associated with black holes, but it is possible that they can exist independent of them. Scientists have, in fact, shown that wormholes may be even more amazing than black holes. It may be possible in the distant future to pass through these wormholes. And if we could, we may be able to traverse extremely long distances in *space* and *time* almost instantaneously. In essence, our age-old dream of flight to the stars and of time machines would be possible. We know it will be difficult to traverse wormholes; nevertheless, it is a challenge that will continue to confront us for years to come.

Fig. 2: A spiral galaxy. (Photo courtesy of the National Optical Astronomy Observatories.)

Not only do Einstein's theories predict strange objects but they have opened new doors. Our present theory of the universe, known as the big bang theory, is based on general relativity. And through the big bang theory, we have been able to trace the universe back to its earliest moments and to study its creation in minute detail. Furthermore, we have been able to predict the fate of the universe. Indeed, as we will see, all of cosmology, from the structure of the universe to the events that occurred just after its creation, depends on general relativity.

Another important discovery made as a result of general relativity came in 1918, when Einstein published papers predicting that oscillating matter would create "gravitational waves," just as oscillating charges create electromagnetic waves. Furthermore, like electromagnetic waves, these waves would travel at the speed of light. Compared to electromagnetic waves, however, they would be exceedingly weak and therefore difficult to detect. Despite the difficulties, Joseph Weber of the University of Maryland searched for them and in 1969 announced that he had discovered them. Scientists around the world were jubilant; if true, it would be a monumental breakthrough. But within a few years it was evident that his results could not be verified. Despite the disappointment, Weber's work was not in vain; he ignited a spark that eventually gave rise to a new branch of astronomy: gravitational wave astronomy. Gravitational wave astronomy is now an important branch of science, and several large projects are now under way in an effort to detect the waves. The major one in the United States is referred to as LIGO (Laser Interferometer Gravitational-Wave Observatory). In other parts of the world several similar projects have been initiated; they are referred to as TAMA-300, GEO-600, and VIRGO. It seems it is only a matter of time before these waves are detected.

In 1936 Einstein made another important prediction that, at the time, he thought would be only of academic interest. He imagined two stars, one directly behind the other, and considered the rays of light from the more distant of the two objects. According to general relativity, the gravitational field of the closer star would bend them. Indeed, the situation would be very similar to the rays of light passing through a lens. But a lens magnifies an object. According to Einstein, the nearby star would do the same thing, and if such a situation occurred, we would see a magnified image of the distant star with the tiny image of the nearby star at its center. We now refer to the phenomenon as "gravitational lensing."[5]

In 1937 Fritz Zwicky of Caltech generalized the prediction to aligned galaxies. He noted that the effect would be much greater in this case. Few paid any attention to the prediction until 1979 when British astronomer

Dennis Walsh discovered a pair of identical quasars (extremely distant galaxy-like objects). He realized that if the nearby object was not exactly in line with the distant quasar, two or perhaps three slightly distorted images of it would occur, rather than a single magnified image. He predicted that it might be a gravitational lense. In short, he predicted that the two images were, in reality, the same object. And this was soon shown to be the case.

General relativity is now considered to be one of the two major pillars of modern physics. It explains the very large. But there is another pillar, namely, quantum mechanics, that explains the very small. Although Einstein made significant contributions to early quantum theory, he was not directly involved in the major discoveries related to the formulation of quantum mechanics. Nevertheless, he played an important role as a critic. With his tremendous ability to penetrate to the heart of a problem, he was able to see things that others could not see, and they troubled him. Many weird things appeared to be predicted, particularly after the Danish physicist Niels Bohr formulated his "Copenhagen interpretation" of quantum theory. Einstein had no qualms about the physical predictions of the theory; there was no doubt that they were accurate, but its philosophical implications bothered him deeply. He struggled to show that the theory had shortcomings. Although he was unable to prove his case, his endeavors shed light on many perplexing difficulties of the theory, and his contributions had a major impact on the field.

For the most part, Einstein was proud of his contributions to science. The success and adulation it brought him, however, puzzled and confused him. Still, he took it all in stride. There was, however, one contribution that caused him considerable discomfort, particularly in his latter years. His equation $E = mc^2$ became the basis of the atomic bomb because it showed that matter and energy were equivalent and that it might be possible in the future to convert mass directly to energy. Indeed, it was soon shown that mass could be converted to energy in certain nuclear reactions, and the implication was immediate: a bomb that would be capable of tremendous explosive power could be built. Einstein realized this but was sure it would not occur in his lifetime. Nevertheless, when he heard, just before World War II, that Nazi Germany was working on such a bomb, he wrote a letter (actually he signed a letter that had already been written) to President Franklin D. Roosevelt urging him to initiate an atomic bomb project. The bomb was, of course, built and used, but not on Germany.

As a strong pacifist, Einstein regretted writing the letter, but he knew the project would have gone ahead without him. He did everything he could to dissuade the United States from using it, and he worried about what would happen if several nations got ahold of such a bomb. He pressed for the formation of a world government to avoid such dangers, but most of his efforts were in vain.[6]

Einstein's most celebrated contributions to science were, of course, his special and general theories of relativity, but he made many contributions that did not receive the same attention yet were of paramount significance to science. Interestingly, he did not receive the Nobel Prize for either of his relativity theories but rather for his explanation of the photoelectric effect. It was in this paper that he introduced the idea that light was composed of quantum particles that we now refer to as photons. In the same year, 1905, he published a paper proving the existence of atoms and molecules and a paper explaining the phenomenon called Brownian movement. Furthermore, in the same year that he published his general theory of relativity, 1916, he published a paper on "quantum jumps" between energy levels. He showed that as electrons jumped they would emit or absorb photons. It was here that he introduced the idea of "stimulated emission," which eventually led to the invention of the maser and the laser.

Another breakthrough came in 1924 when Einstein published a paper based on a letter he received from the Indian physicist Saryendra Bose. The paper led to the formulation of a new type of statistics, known as Bose-Einstein statistics, which is an important tool in modern physics. As a result of this paper, Einstein later made a remarkable prediction concerning the behavior of matter at extremely low temperatures. He predicted what is now called "superfluidity," a phenomenon we will look at later.[7]

There's no doubt that Einstein was extremely productive during his youth, but many people felt that he lost touch with mainstream physics as he grew older. He spent the last thirty years of his life in a quest that many scientists considered foolish: a search for a unified field theory—a theory that would unify the gravitational and electromagnetic fields. He hoped this theory would explain the elementary particles that were known at the time. In short, what he really wanted was a theory of everything. In this endeavor he was outside the mainstream of physics, and most scientists frowned on his efforts.

Einstein's major problem in trying to formulate a unified field theory was quantum mechanics. Quantum mechanics was well established, and it appeared to solve all the major problems of the microworld. Einstein was going against the tide by trying to formulate a unified field theory by extending his general theory of relativity. There was a serious problem in taking this approach: general relativity was not a quantum theory. He hoped that quantum mechanics would somehow be incorporated in his extension of general relativity. But after years of futile attempts, it appeared as if there was no way the two theories would come together.

Most people looked upon his endeavor as a failure, but in reality his work eventually encouraged others to look for a unified theory and also a theory of everything. Most of these efforts were based on quantum theory, however, rather than general relativity. Four fields of nature were known. Besides the gravitational and electromagnetic fields, there were the strong and weak nuclear fields. Three of these fields were eventually unified (tentatively) in what is called the grand unified theory (GUTs, for short). But gravity remained elusive. It was soon evident that desperate measures were needed to bring gravity into the fold. The most successful of these efforts is known as superstring theory.[8]

Superstring theory is a new and entirely different approach to the problem. According to the theory, fields and elementary particles are made up of tiny strings that are trillions of times smaller than atoms. To further complicate things, these strings exist in higher dimensions. Hermann Minkowski of Göttingen extended Einstein's special theory, shortly after it was formulated, to four dimensions by assuming time was the fourth dimension. And Theodor Kaluza of Königsberg extended general relativity to five dimensions in an attempt to incorporate the electromagnetic field. With superstring theory, scientists have gone beyond this to ten dimensions and even more. It has had a number of successes, but difficulties persist. It is now evident that something beyond superstring theory is needed. Because of this, M-theory was developed.

Einstein's theories have, indeed, had a far-reaching and profound influence on modern physics. They have helped shape many areas and have led to several new branches of physics. In the following pages we will look into these startling new areas and ponder Einstein's vision of the universe.

Chapter 1

Twists in the Fabric of Space

In the mid-1600s Sir Isaac Newton formulated the fundamental laws of motion and a theory of gravitation. One of his major postulates was that space and time were absolute; in other words, they were the same for all observers, regardless of their motion or position in the universe. Two hundred and fifty years would pass before someone would finally realize that this wasn't correct. Albert Einstein puzzled over this problem for several years, and in 1905 he showed in his special theory of relativity that both space and time depended on the motion of the observer. Using rather simple mathematics, he proved that measuring rods moving relative to us would appear contracted in length and that clocks moving relative to us would appear to run slow. The effects in both cases would be small at everyday velocities but would be significant near the speed of light. One of the strangest results of the theory was its prediction that measuring rods would contract to nothing and clocks would stop at the speed of light (relative to us). This indicated that the speed of light had to be an uppermost speed for matter in the universe. In essence, matter could move at any speed up to the speed of light relative to us but not *at* the speed of light.

The basic equations of space and time in special relativity dealt only with uniform, or nonvarying, straight-line motion. Furthermore, Einstein soon realized that it said nothing about gravity. Newton's theory of gravity was still the accepted theory, even though, as Einstein soon showed, major conflicts existed between it and special relativity. One of the conflicts was that the effects of gravity were instantaneous according to Newton. If a mass suddenly appeared at some point in space, the gravitational field around it appeared instantaneously, which meant that the gravitational field had an infinite speed. But according to special relativity, nothing could travel at speeds greater than that of light.

And this wasn't the only problem. According to Newton, the gravitational attraction between two objects in space depended on their respective masses and on the distance between them. According to special relativity, however, when we are dealing with two objects, the mass of each of the objects depends on its velocity with respect to the other object. Furthermore, the distance between the two objects also depends on their relative motion. These were serious problems, and they seemed to indicate that Newton's theory was only an approximation to a more exact theory.

THE YEARBOOK ARTICLE

In 1907, two years after Einstein completed his special theory of relativity, Johannes Stark, the editor of the *Yearbook of Radioactivity and Electronics*, asked him to write a review article on relativity for the yearbook. Einstein considered it an honor and gladly accepted; it would, in fact, be a good opportunity for him to go through everything he had done and speculate on how the theory might change in the future. He was working at the patent office in Bern at this time and had already been thinking about how the theory should be extended; furthermore, he knew there were problems between it and Newton's gravitational theory. As he thought about how the problems could be overcome, he realized there was a link between relativity and gravity. He pictured someone falling from the roof of a high building. It was immediately obvious to him that the person would not feel his weight; in essence he would be weightless. Using this, he quickly arrived at another fundamental insight. Acceleration produced a force on a person's body similar to the one gravity produces, and there-

Fig. 3: Einstein.

fore acceleration and gravity must be related. Einstein later referred to this as the "happiest thought of his life."[1] He was overjoyed and soon realized it was exactly what he was looking for to solve his problem.

It was now obvious to him why special relativity didn't apply to gravity. Gravity was associated with acceleration, and the space and time equations of special relativity dealt only with uniform, straight-line motion. The force a person felt when he was being accelerated was known as an inertial force (the force that resists a change in motion), and Einstein now knew that inertial forces and gravitational forces were connected.

As it turned out, Einstein wasn't the first to consider the similarity between inertial forces and the force of gravity. It had been known for years that the mass (weight) of an object could be determined in two ways. You could step on a scale and determine your weight (mass); this is referred to as your gravitational mass. But if no gravitational field is present, you could also determine your mass by giving yourself a push. Newton's second law of motion tells us that the greater the mass, the greater the force needed to set it in motion, or more exactly, to give it a particular acceleration. In theory, you could push on a mass and measure its acceleration; this would give you its inertial mass.

Newton and others knew that gravitational mass and inertial mass were equal. But what surprised them was how *exactly* equal they were. In 1889 Roland von Eötvös of Hungary showed that they were the same to one part in a hundred thousand. In 1964 Robert Dicke of Princeton University took this even further; he showed that they were equal to one part in a hundred billion. Newton's theories and those of others could not explain this. It was just a "coincidence" to them.

Why, then, was Einstein's "happiest thought" so important? Because Einstein assumed that not only were they numerically equal, but they were more than this. There was no difference between a "real" gravitational field and one produced via inertia. Let's look a little closer at this. Consider a man in an elevator with no windows. If the cable holding the elevator in place suddenly broke, he would be in a free fall and would not feel his own weight. With space rockets and the space station this is, of course, not new to us; we frequently see astronauts on TV and in the movies floating around freely in gravity-free surroundings.

Let's assume now that the elevator is out in space with no gravitational field nearby. Assume also that it is pulled upward with an acceleration equal to the acceleration of gravity on Earth (32 ft/sec^2). The man in the elevator would be pressed against the floor with a force equal to that of the gravitational pull on Earth. Indeed, if there were no windows in the elevator, he would likely think he was on Earth. It might seem that we are creating an "artificial" gravitational field by doing this. But Einstein said this isn't the case. He said that we are, indeed, creating a "real" gravitational field. In short, it is just as real as the gravitational field on Earth. There is no difference between the two cases, and no matter how many different experiments we performed within the elevator, there is no way we could tell the difference. (Actually, as we will see later, there is a slight difference, but we'll ignore it for now.)

THE EQUIVALENCE PRINCIPLE

Einstein called his idea the *equivalence principle* since it implied that gravitational mass and inertial mass were equivalent and also that gravity and acceleration were equivalent. Within a short time, Einstein was able to use the principle to make several predictions. He showed that clocks would run slower in a strong gravitational field, as compared to a weak one. In other words, a clock on the surface of Earth would run slower than one in space. The effect in the case of Earth would be extremely small, but in places where the gravitational field was much stronger than Earth's, the effect could be significant. Clocks on the surface of the Sun, for example, would run slightly slower than those on Earth, and the effect would be much greater on very dense, massive stars.

Einstein also used his equivalence principle to show that the path of light rays would be affected by a gravitational field. They would be deflected from straight-line motion; and the stronger the field, the greater the deflection. The reason for this is easy to see if we go back to our accelerated elevator. Assume first of all that the elevator is sitting out in space. If we drill a small hole in one side and let a beam of light from a star pass through the elevator, it would cross from one side to the other in a straight line. We could easily mark the spot where it hits the other side of the elevator. Now assume that a cable is attached to the top of our elevator and that it is accelerated upward (figure 4). In this case, as the beam moves across the elevator, it will be deflected downward. But, as we just saw, acceleration is equivalent to gravitation, so a gravitational field would also deflect the beam of light.

Einstein wrote up these results for his yearbook article, and it was

Fig. 4: An illustration of the equivalence principle. The downward accelera-tion of the elevator is the same as the acceleration of gravity on Earth.

published in early 1908. Although it contained some important break-throughs, hardly anyone noticed it, and no one, with the exception of Max Planck of the University of Berlin, commented on it. Einstein continued working on the problem for a while, but he eventually became discour-aged and soon had little time for research. After several frustrating years he finally managed to land a professorship at the University of Zurich, and lecture preparation began to take up much of his time. In addition, his attention was being taken up more and more by quantum theory.

TO PRAGUE

Einstein did not stay long at the University of Zurich. In 1910 the German University of Prague began scouting him, and in January 1911 they made him an offer. His teaching load at the University of Zurich was heavy, and

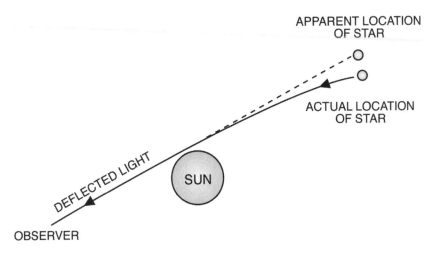

Fig. 5: The deflection of a beam of light by the Sun.

it allowed little time for research. At Prague he would have more time for research, and his wages would be considerably higher. So in March 1911 Einstein and his family moved to Prague.[2] He enjoyed his new position at first but soon became dissatisfied with Prague and its people. However, he was now able to get back to his theory of gravitation, and he began thinking about his principle of equivalence. Within a short time he determined that a beam of light that grazed the limb of the Sun would undergo a deflection of .83 second of arc (figure 5). Certain that this could be measured, he approached the astronomy department at Prague. One of the astronomers, Leo Pollack, was traveling to Berlin, and he visited a young astronomer, Erwin Freundlich, at the Berlin Observatory. Freundlich took an immediate interest and contacted Einstein, offering to help. Einstein was disappointed to learn, however, that the next eclipse wasn't until September of 1914, and he didn't want to wait that long. Freundlich offered to check previous eclipse photos at the Hamburg Observatory to see if any stars were visible near the Sun, but to his disappointment, none of the plates were usable. Einstein then suggested that the rays deflected by Jupiter could be measured, but when the calculation was made, the effect was found to be far too small. As a last resort, Einstein wrote to George Hale at the Mt. Wilson Observatory in California, asking if it was possible to observe stars near the limb of the Sun without an eclipse. Hale replied that it wasn't.

So far Einstein had been dealing only with the predictions of the theory. He now turned his attention to the formulation of the theory itself. The equations of special relativity would obviously have to be generalized. In his first attempt he assumed that the speed of light was variable, and in June 1911 he submitted a paper to *Annalen der Physik* presenting a new theory. As it turned out, Einstein wasn't the only one working on a theory of gravity. Max Abraham of Göttingen had also published a theory, and he was soon attacking Einstein's theory. Einstein, on the other hand, had little use for Abraham's theory. The result was several articles in *Annalen der Physik* attacking each other's theories. The arguments became quite heated, particularly on the part of Abraham, who was incensed that Einstein dismissed his theory as insignificant—with little chance of being correct. In private, Einstein described it as an "embarrassing monstrosity."[3] Most of Abraham's criticisms of Einstein's theory centered on his variation in the speed of light, and eventually this aspect of the theory began to worry Einstein.

About the same time another theory of gravitation with a constant speed of light was put forward by Gunnar Nordström of Finland. Einstein corresponded with Nordström, and this time the discussions were much more friendly. A little later Einstein published a paper with a Dutch physicist, Adrian Fokker, in which they discussed Nordström's theory. In particular, they compared Einstein's theory to Nordström's and, as you might expect, came to the conclusion that Einstein's theory was more acceptable.

THE ROTATING DISK PARADOX

Einstein soon realized that there was a serious problem with his variable speed of light theory. It was based on Euclidean geometry, and there were indications that the geometry of gravitation was not Euclidean. These indications came from the problem of the rigid rotating disk (see figure 6). When special relativity was applied to this disk, the results were hard to explain. Each small arc along the circumference would contract according to special relativity, but the radius, which is perpendicular to the circumference, would not change. This meant that one of the pillars

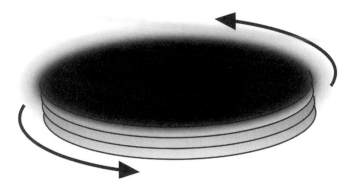

Fig. 6: A rotating disk. Einstein showed that there was a paradox associated with the rim of the rotating disk.

of Euclidean geometry, namely, that the ratio of the circumference of a circle to its radius was π (3.1415), could not be true of a disk that was rotating. Special relativity implied that, because of its shrunken circumference, the disk would break up at high speed. Einstein spent considerable time discussing this problem with his close friend Paul Ehrenfest, and Ehrenfest published several papers on it. It eventually became known as the *Ehrenfest paradox*.[4]

What was critical about the paradox was that when the disk was rotating, the circumference was accelerating, and according to the principle of equivalence, acceleration and gravity were the same. When the disk was not rotating, therefore, it was as if there were no gravitational field. But when it began to rotate, a gravitational field was generated, and the ratio of the circumference to the radius would change (and it would not be π). To Einstein, this signaled that Euclidean geometry was no longer valid in a gravitational field.

Why is this important? The main reason is because Euclidean geometry is the basic geometry of our three-dimensional world; it's the geometry you likely took in high school. It is based on five axioms, or self-evident truths. For years mathematicians puzzled over the fifth axiom, which didn't seem as fundamental as the other four. It is as follows: Assume we have a point off to the side of a straight line. Through this point only one line parallel to the existing line can be drawn. The German mathematician Karl Gauss saw a flaw in this, noticing that on a curved surface it was no longer true. His idea was picked up by the German mathematician Bernhard Riemann and others, and eventually two different non-Euclidean geometries were developed.

Einstein remembered a class he had taken at Zurich Polytechnic that was taught by Carl Geisser. Geisser had covered Gauss's theory of surfaces, but Einstein didn't remember much about it. The more he thought about it, however, the more he realized that his new theory would have to be non-Euclidean. He talked to Georg Pick of the mathematics department in Prague, and Pick pointed out that considerable work had been done in the area of non-Euclidean geometry. Einstein does not appear to have followed up on his suggestion, however.

About this time Einstein also came to another important decision. His old teacher at the Zurich Polytechnic, Hermann Minkowski, had published a new formulation of special relativity. He had brought space and time together as four-dimensional "spacetime." Most scientists considered it an elegant and beautiful modification, but Einstein didn't like it. "Since the mathematicians have got ahold of my theory, I hardly recognize it,"[5] he said. He was sure such elegance wasn't needed. But as the new formulation attracted more and more attention, with almost everyone applauding it, Einstein began to look at it more closely, and he soon realized it could play an important role in the formulation of his new theory.

Einstein came to another conclusion at about this time (or perhaps a little earlier). He noticed that the equivalence principle was restricted; it was strictly valid only in an infinitely small space. To see why, consider the "lines" of gravity around Earth. They radiate from its center as shown in figure 7. We know that when the lines are parallel, the field is uniform, and when they are closer together, the field is stronger. Above Earth they are obviously not parallel; they get closer together as they approach Earth. If we consider an accelerating elevator, on the other hand, these lines are parallel and uniform. Comparing an elevator above Earth to one accelerated out in space, we therefore see a slight difference. This difference gives rise to what are called "tidal forces."

Let's look at the significance of these tidal forces. We'll begin by taking a closer look at the gravitational field lines in our elevator. It's important to remember that the closer the lines are together, the stronger the gravitational field. This means that when the elevator is in the gravitational field of Earth, the field is stronger at the bottom of the elevator

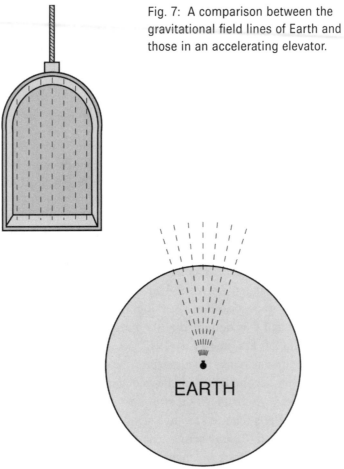

Fig. 7: A comparison between the gravitational field lines of Earth and those in an accelerating elevator.

EARTH

because the field lines are closer together there. Thus, if we were out in space and dropped two balls, one from a height above Earth higher than the other, the lower one would accelerate toward Earth at a greater rate than the higher one. As a result, the separation between the two balls would increase as they approached Earth's surface.

How would this affect an object or a person, say, an astronaut, falling toward Earth? It's easy to see that his legs would be pulled with a greater force than his head, and as a result, his body would be stretched. Interestingly, he would also feel another force. Looking back at our example of falling balls, and assuming this time that they are falling side by side, we know that they will follow the lines of gravity. But these lines get closer together as they approach Earth; therefore, the two balls will be forced together. This means the astronaut will not only be stretched, but he will also be squeezed. Einstein got around this problem in his principle of equivalence by assuming the elevator was very small. To a good approximation, the lines of gravity would be parallel in this case.

BACK TO SWITZERLAND

Einstein made considerable progress on his new theory while in Prague, but he soon hit a barrier. He knew he would have to use more complicated mathematics to set it up—a four-dimensional generalization of Gauss's theory of surfaces—and he wasn't sure how to proceed. Aside from Pick, there was no one to talk to in Prague about his problems, but this situation was soon to change. His friend Marcel Grossmann was now a dean at his old alma mater, Zurich Polytechnic, and he offered Einstein a job. Other offers were also now starting to come in, but Einstein loved Switzerland and was delighted at the prospect of returning. He accepted the offer and left Prague in July 1912.

An additional bonus awaited Einstein in Zurich. His friend Grossmann was an accomplished mathematician and had done his thesis on non-Euclidean geometry. Soon after arriving in Zurich, Einstein went to Grossmann. He is reported to have said, "Grossmann, you've got to help me or I'll go mad."[6] He described his problem to Grossmann, and Grossmann immediately took an interest. To Einstein's delight, he agreed to help him, but only on the condition that he have nothing to do with the physics; he would help only with the mathematics. Einstein quickly agreed. Although he had done a thesis on non-Euclidean geometry, Grossmann's knowledge of the area was limited. He went to the library, did the necessary research, and soon found that considerable work had been done by Bernard Riemann, Gregorio Ricci, and Tullio Levi-Cevita. They had developed what is known as *tensor analysis*, a very difficult and complex branch of mathematics. But Grossman could see that it would be ideal for Einstein's problem—exactly what he needed.

To see what a tensor is, let's begin with what is called a vector. A quantity such as velocity, or acceleration, is a vector because it has both magnitude and direction. For example, when you are traveling in a car,

Fig. 8: Marcel Grossmann.

you have both a speed and a direction; the combination of these two is the velocity of the car. In three-dimensional space a vector such as velocity has three components, one along each of the three dimensions of space (we frequently refer to them as the x, y, and z directions). A tensor is a generalized vector in that it has more components (actually, a vector is a simple tensor). Grossman soon saw that a particular tensor discovered by Riemann, called the curvature tensor, would be useful. It gave a measure of the curvature of a surface or a space, and it could be used to describe the gravitational field. By now Einstein had come to the conclusion that gravity was curved space. To many, this seemed a little weird. Einstein knew that no one could see the "curvature" of space, but it could be dealt with mathematically, and that was all that mattered to him. There was, however, a drawback. In Newton's theory only one component was needed to describe gravity; with Riemann's tensor, ten components would be needed. So the theory was obviously going to be much more complicated than Newton's.

Another problem was related to the form of the theory. It was important, as far as Einstein was concerned, that his new theory be independent of the coordinate system. In other words, the laws of nature had to be the same when you switched from one coordinate, or reference, system to another; for example, if you switched from Cartesian (x, y, z) to curvilinear coordinates (r, θ, ϕ). This is referred to as covariance. Without covariance, it would be like measuring a rod with a yardstick (in inches), then measuring it with a meter stick (in centimeters) and getting two different answers when you converted inches to centimeters. Grossmann soon found a slight variation of the Riemann tensor that would give a covariant theory. It was called the Ricci tensor. The two men also identified a tensor that could represent the mass of an object; it was called the mass-energy tensor.

Einstein was pleased that things were coming together so quickly. They had a good candidate for the equations of the gravitational field, and they appeared to be covariant. These equations would give the form of the gravitational field around any massive object, or more exactly, the shape of the space around any massive object. It was well known, however, that Newton's theory of gravity gave excellent predictions for relatively weak gravitational fields. This meant that in the limit of weak fields, Einstein's theory had to go over to Newton's theory. In other words, they would have to give the same answer. Indeed, the only known deviation from Newton's theory was a slight discrepancy in Mercury's orbit. The position of Mercury could not be predicted exactly by Newton's equations, but the discrepancy was small.

When Einstein and Grossmann checked their new equations, however, they were disappointed. They did not appear to give Newton's equations. A further disappointment came when they looked for other covariant equations. There didn't appear to be any; the Ricci tensor was the only tensor that gave covariant equations. Reluctantly, Einstein began looking at possible noncovariant equations, and finally he and Grossmann arrived at one that appeared to be satisfactory. They published their new theory in early 1913. Einstein had misgivings about the theory; it was not covariant, and this bothered him. He referred to the noncovariance as an "ugly dark spot" on the theory.[7] But there seemed to be no alternatives; Einstein could find no equations that were better. The paper was titled "Outline of a Generalized Theory of Relativity and a Theory of Gravitation." The German word for outline is *entwurf*, and soon the theory was referred to as the Entwurf theory.

Einstein didn't like the noncovariance of the theory, so he set out to show that a gravitational theory based on curved space could not be covariant. And within a short time he had two proofs—or at least what he thought were proofs. The second one was much stronger than the first and was soon referred to as the "hole" proof. According to it, a covariant gravitational theory would not satisfy "causality." According to causality, every effect must have a cause, and the cause must come first. In the case of a baseball that is struck by a bat, the ball must be thrown by the pitcher before it can be hit by the bat. It's not possible that the ball could fly over the center field fence, then a few seconds later the pitcher throws it. Einstein also showed that covariant equations would not be unique; in other words, they would have more than one solution. These so-called proofs would later be an embarrassment to him.

TO BERLIN

Einstein's stay at the polytechnic in Zurich was short. His fame was now spreading rapidly, and the University of Berlin, one of the largest and most prestigious universities in Europe, soon became interested in him. Max Planck and Walther Nernst of the University of Berlin visited him in Zurich during the summer of 1913, making him an offer he couldn't refuse. Not only would he get a considerable increase in salary, but he wouldn't have to do any teaching; he would be able to spend most of his time doing research. This is the aspect of the offer that appealed to him the most. He had a fairly heavy teaching load at Zurich Polytechnic, and he was starting to resent it. He wanted more time for research. There was a drawback, however; he had a distrust and dislike of Germany. He had, in fact, left Germany many years earlier because of its militarism and strong discipline. In addition, his wife, Mileva, dreaded the thought of a move to Berlin; she loved Zurich and wanted to stay there, but their marriage was now on the rocks, and she had little influence with Einstein. As far as he was concerned, the offer was too good to turn down, and he accepted it.

Einstein arrived in Berlin on April 3, 1914. His collaboration with Grossmann was over, but he was now an expert in tensor analysis and needed little help with the mathematics. Once in Berlin, he devoted himself full time to his research, determined to make his new theory work. But there were still problems. Some of them were pointed out by the Italian mathematician Tullio Levi-Cevita—one of the developers of tensor analysis—when an intense correspondence began between the two men. Their correspondence would last for almost a year. Levi-Cevita's main objection was the way Einstein derived his equations; strangely, he was not concerned with the noncovariance of the theory. Einstein was flattered by his interest. In his reply to Levi-Cevita's first letter, he wrote, "When I saw that you attacked the most important demonstration of my theory, which I obtained with streams of sweat, I became not a little alarmed, particularly because I know that you have a much better mastery of these mathematical matters than I do."[8] Levi-Cevita was not able to convince Einstein, but he did worry him. Einstein tried to counter Levi-Cevita's arguments in his reply, but neither man was able to persuade the other. Nevertheless, the letters had a positive effect on Einstein. He began looking at his theory with a much more critical eye, and it can be said that this exchange was the starting point of Einstein's dissatisfaction with the Entwurf theory.

In June, Einstein went to Göttingen at the invitation of David Hilbert and Felix Klein. At the time, Hilbert was acknowledged as the greatest mathematician in the world. In six two-hour seminars Einstein presented his theory to the Göttingen scientists. He was thrilled by Hilbert's interest and his understanding of it. Writing to Sommerfeld, he said, "[It was] a great pleasure to be understood in every detail. I am quite enthusiastic about

Hilbert. A great man."[9] Like Levi-Cevita, Hilbert also zeroed in on the weak spots of the theory, and as a result, by the time Einstein got back to Berlin, he was becoming increasingly distressed with the theory. He still didn't like its noncovariance, and when he calculated the deviation that it predicted for Mercury's orbit, he got 48 seconds of arc, only about one-third of the observed amount. Then he determined that the theory did not give a satisfactory explanation of the rotating disk. This was too much. He finally decided that the theory had to be wrong. But what was the alternative?

THE BREAKTHROUGH

After considerable thought, Einstein decided to take another look at the covariant theory he and Grossman had developed two years earlier—the one based on the Ricci tensor. He soon found that he had made an error: the theory did give Newton's theory in the weak field approximation. He was overjoyed. He calculated the prediction for Mercury's orbit, and it fit observation exactly. The discrepancy was due to a precession, or slight shifting, of the ellipse of the orbit that was not accounted for in Newton's theory (see figure 9). He was sure now that he was on the right track. The

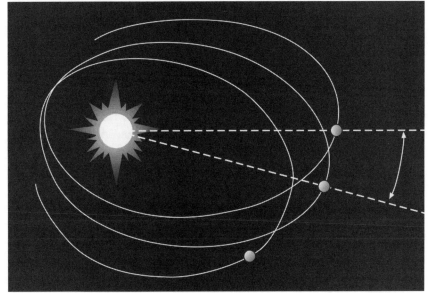

Fig. 9: The precession of Mercury's orbit.

next prediction was the deflection of light rays grazing the limb of the Sun. To his surprise, he got 1.75 seconds of arc, which was twice the amount he had obtained earlier. As it turned out, this discovery was helpful to him, since it was now known that in 1801 the German astronomer Johannes Soldner had made a prediction of .87 second of arc for the deflection, based on Newton's theory. Newton had, in fact, even hinted at a deflection. Einstein's prediction was double theirs, so there would be no doubt when it was finally measured.

Einstein presented his new theory to the Prussian Academy at one-week intervals, beginning on November 4, 1915. In the first session he outlined the difficulties he had encountered in his Entwurf theory and stated that he was abandoning it. His new theory was going to be based on the Ricci tensor. On November 18 he presented his results for Mercury's orbit and showed that they agreed exactly with observation. He was particularly pleased that the numbers agreed so well without having to introduce any "fudge factors," as was the case in many other theories. He also presented his new value for the deflection of a light beam grazing the Sun.

Throughout the first couple of weeks of November, Einstein worked as he had never worked before, but in his November 18 presentation to the Prussian Academy, there was still a problem. The theory was valid for the space around massive objects; it was valid, for example, for the region of the solar system around the Sun, but it was not valid within the Sun. It gave the wrong answer. Einstein continued to work feverishly on the problem. At first he thought the main difficulty was that a proper theory of matter had not been developed, and he decided to work out the details of such a theory. Then he took another look at his equations and discovered that they did not satisfy conservation principles. He added a term so that they would satisfy conservation, and this turned out to be exactly what he needed. With this term, his theory was valid both inside and outside matter. Furthermore, the equations were still covariant. A week later, on November 25, Einstein presented his full theory to the Prussian Academy. It was now complete. Over the next few days he wrote to several of his colleagues and friends expressing his joy and relief at finally finding the correct equations. To Ehrenfest he wrote, "Imagine my joy . . . over the result that the equations correctly yield the perihelion movement of Mercury. For some days I was beside myself with excitement."[10] To Besso he wrote, "My wildest dreams have now come true."[11]

Einstein had barely presented his new theory when, to his dismay, he discovered that his friend Hilbert may have scooped him. Actually, Einstein was more convinced that Hilbert had stolen his ideas and presented them as his own. After Einstein presented his earlier Entwurf theory at Göttingen, Hilbert became particularly interested in it and began his own search for covariant equations. Like Einstein, he soon narrowed in on the Ricci tensor, and just a few days before Einstein presented his complete equations to the Prussian Academy, Hilbert presented his equations to the Göttingen Academy. The proceedings of the Götttingen Academy were published in March, and Hilbert also had the additional term in his equations.

Did he actually scoop Einstein? He may have worked independently, but he had access to almost all of Einstein's work. Besides spending a week at Göttingen outlining his theory, Einstein kept up a steady correspondence with Hilbert while he was developing his new equations in the late fall. In addition, John Stachel of Boston University found that Hilbert had made changes in his manuscript while it was in press during December.[12] He added the extra term that made the equation satisfy conservation. Indeed, it appears as if he saw the term in Einstein's equation and changed his before the paper went to press in March.

Einstein was quite upset at first, sure that Hilbert was trying to steal his ideas. For several weeks he did not correspond with him, but after Hilbert apologized and acknowledged Einstein as the discoverer of general relativity, the two men resumed their friendship.

With all the stress and pressure Einstein was under during the fall of 1915, it's amazing he was able to make the breakthrough. He was under constant stress, not only from the difficulties of formulating his new theory but also from his personal life. His wife had just left him and returned to Zurich, taking his two boys. The marriage had deteriorated so much that the loss of his wife was not a problem; he was particularly close to his boys, however, and the loss of them was a shock to him. In addition, Germany was now in the middle of a war, and as a strong pacifist, he was affected. And finally, he had hoped his theory would be verified experimentally—particularly the deflection of a light beam grazing the Sun—and every attempt had been thwarted. Freundlich and several other astronomers journeyed to Russia for the eclipse of 1914, but they were taken prisoner when World War I broke out. Fortunately, a few weeks later they were traded back to Germany for some Russian officers.

It is, without a doubt, a credit to Einstein's perseverance that he was able to put everything aside and concentrate on the task at hand.

THE WARPING OF SPACE

Einstein's new theory predicted that gravity was a curving or warping of space. The closer you got to the mass, the greater the curvature. A planet followed the curvature, traveling in what is called a geodesic, which is defined as the shortest distance between two points. In flat space this shortest distance is, of course, a straight line, but in curved space it is a curve. The easiest way to visualize it is to use an analogy. Let's assume the Sun is a sphere about the size of a basketball, and assume we place it on a stretched rubber sheet. It will, of course, indent the sheet. If we now take a small sphere such as a marble and project it around the Sun sphere, it will take up a path similar to the orbit of a planet (figure 10).

Einstein did not find an exact solution to his equations; his solution was an approximate one. To his delight, however, within a short time someone did find an exact solution. Einstein received a letter from Karl Schwarzschild, an astronomer with an excellent mathematical background, who was on the Russian front. Schwarzschild was intrigued with Einstein's theory when he first saw it and was soon looking for a solution to the equations. He looked at the region around matter first, and soon found a solution, which he sent to Einstein.

Einstein was surprised that an exact solution was found so soon. He wrote back to Schwarzschild, "I had not expected that one could formu-

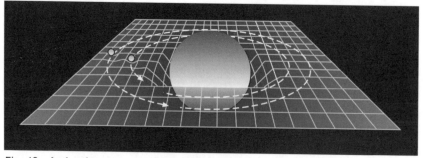

Fig. 10: A simple representation of the curvature around the Sun. Planets follow this curvature in their orbits.

late the exact solution of the problem in such a simple way."[13] A little later he received a second letter from Schwarzschild. He had now solved the equations for the case inside matter. Einstein presented the two solutions to the Prussian Academy in early 1916. Schwarzschild, unfortunately, did not live to see the importance of his results. He contracted a rare disease on the Russian front and died in May at the age of forty-one.

THE *ANNALEN DER PHYSIK* PAPER

Einstein's paper on general relativity was published in *Annalen der Physik* in early 1916 with the title "The Foundations of the General Theory of Relativity."[14] It was one of the longest papers he had ever published. After outlining the shortcomings of special relativity and Newton's theory of gravity, and suggesting that a new theory was needed, he began with a long section on the mathematics of tensors. Most physicists were not familiar with tensors at that time, and he no doubt felt that some discussion of them was needed. He described the properties of tensors and introduced the Riemann and Ricci curvature tensors. Then he introduced his equations for the gravitational field, considering the case outside matter first, then introducing the more general case that included matter. He went on to show how his equations would affect Maxwell's electromagnetic equations.

It was important, of course, to show that his theory gave Newton's theory in a first approximation, and he did that in the next section. He then went on to show that clocks would run slow in an increased gravitational field, and he derived the new expression for Mercury's orbit, showing that it agreed exactly with observation. Einstein was pleased with his theory, and as we will see in the next chapter, it helped us understand our universe better.

Chapter 2

Expanding to Space

History of the
Cosmological Constant

W hen Einstein finished his general theory of relativity in 1916, he wanted to apply it to as many problems as possible, and it was soon obvious to him that one of the most exciting applications would be to the universe. What would the theory tell us about the universe? The only theory of this type that had been formulated at the time was based on Newton's theory of gravity, and it had encountered serious problems. Newton had to decide whether the universe was finite or infinite, and as it turned out, both cases presented difficulties. If the universe was finite, objects such as stars within it would be attracted to one another, and the universe would eventually collapse. The alternative, an infinite universe, also posed problems. Infinities were difficult to visualize, and a universe that went on forever was not satisfactory to most scientists. Furthermore, it was eventually shown that even an infinite universe would eventually collapse on itself.

The concept of a boundary was obviously a serious difficulty. First of all, how could you give the universe boundary conditions? Furthermore, if it had a boundary, someone would be sure to ask: what's on the other side of the boundary? These difficulties were more than Newton could handle, and he never did formulate a satisfactory theory of the universe. Einstein knew he would have to face the same problems, but he had a new and powerful tool, namely, general relativity. He began by assuming that the universe was homogeneous throughout and isotropic (the same in all directions). And after considerable thought he decided that the universe would have to be finite. Infinities were too difficult to deal with. He also wanted his universe to satisfy Ernst Mach's principle. This principle is concerned with the relation between inertia and the matter of the universe. It's best illustrated with an example. Suppose you were out in space and there were no massive objects in the space. Would you feel an inertial force when you pushed on something? Mach said no, and his idea eventually became known as Mach's principle. Einstein was convinced that Mach was right, and he hoped to incorporate the principle into his cosmology.

Einstein spent a lot of time thinking about the difficulties of formulating a theory of the universe. At one point in his famous 1917 paper on cosmology, he wrote, "I shall conduct the reader over the road that I have myself traveled, rather a rough and winding road, because otherwise I cannot hope that he will take much interest in the result at the end of the journey."[1] Once he had decided on the type of universe he wanted, Einstein applied his theory, but to his dismay he found that it predicted an unstable universe. It would either collapse on itself or expand.

Einstein knew he had to do something. With great reluctance, he changed his field equations, adding a term he called the cosmological constant (it was actually multiplied by another term). He felt that the additional term destroyed the simplicity and beauty of his equations, but it affected things only on a very large scale and did not change anything on the scale of the solar system. Furthermore, the equation was still covariant (the form of the equation did not change when the coordinate system was changed), and energy was still conserved.

Einstein's model of the universe was non-Euclidean in that it didn't satisfy the requirements of Euclidean geometry. It was the three-dimensional analogue of the surface of a ball. Scientists refer to it as a spherical universe, but there is another dimension, namely, time, and it is not curved. Including time, the universe is more like a cylinder in four dimensions. The space part is still spherical, however, and we know that a spherical surface is finite, so Einstein's universe was finite, and he didn't have to worry about satisfying boundary conditions. Furthermore, its spherical shape came about naturally; it was created by the matter of the universe. As we saw earlier, matter curves space. If you shone a light beam into Einstein's universe, it would be bent by the matter in it and would trace out a huge circle in the universe, eventually arriving back at the point from which it began.

Little was known about the universe when Einstein formulated his model. We now know that it is filled with galaxies, or "island universes" of stars, but the idea of a galaxy was unknown to Einstein (although Immanuel Kant had discussed the possibility), and he assumed that the universe was populated with stars. He thought of them as molecules in a gas. They could have random motions, but overall the universe was static. In his paper he wrote, "The most important fact we draw from experience as to the distribution of matter is that the relative velocities of the stars are very small as compared with the velocity of light. . . . There is a system of reference relative to which matter may be looked upon as being permanently at rest."[2]

With his model, Einstein was able to make some important predictions. He showed that the cosmological constant was related to the radius of the universe and could be determined from the average density of the universe. This meant that if you knew the average density, you could determine the size of the universe. Einstein did not make any predictions in the paper because he didn't know the average density. But by March 1917 he had obtained an estimate of the density from Edwin Hubble in the United States. Substituting it into his equations, he obtained a radius of 10^7 light-years. He did not publish the result but related it in a letter to the Dutch astronomer Willem de Sitter. He mentioned to de Sitter that astronomers were only seeing out to about 10^4 light-years, so the estimate seemed to him to be reasonable.

DE SITTER'S UNIVERSE

Only a few months after Einstein published his cosmology, de Sitter published his own cosmology. As the Dutch foreign correspondent to the Royal Astronomical Society in England, de Sitter was one of the few who had access to Einstein's theory during World War I. De Sitter passed on

Fig. 11: Willem de Sitter.

information about Einstein's new theory to Arthur Eddington in England, and Eddington soon became intrigued with it. He asked de Sitter to write several papers on it for the *Monthly Notices of the Royal Astronomical Society*. While writing up his third contribution, de Sitter looked carefully at Einstein's cosmological model, and to his surprise he noticed that Einstein had missed a solution. De Sitter developed the solution and published it in the article.

De Sittter's model was strange, to say the least. First of all, it was empty in that it had no matter in it. It was static, as Einstein's model was, but with no matter in it, you had to ask: what was static? Indeed, within a short time Eddington and Hermann Weyl showed that if two bits of matter were placed in de Sitter's universe, they would repel one another. Furthermore, de Sitter had shown that distant stars, or galaxies, would exhibit a shift in their spectral lines toward the red end of the spectrum due to the Doppler effect, but he thought it was spurious.

To many, an empty universe didn't make much sense. De Sitter coun-

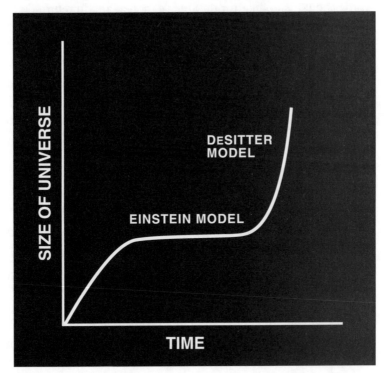

Fig. 12: The expansion of the Einstein and de Sitter models of the universe.

tered this by stating that the observed universe is very close to being empty, and indeed it is. The average density of the universe is only about 10^{-28} gms/cm^3. De Sitter also wasn't sure what to make of the prediction that objects within his universe expanded away from one another. There was no evidence that the universe was expanding.

Soon after formulating his theory, de Sitter sent a copy of his paper to Einstein. Einstein's reaction, as you might expect, was negative. He was embarrassed that he had missed the solution, but an empty universe didn't appeal to him. First of all, it contradicted Mach's principle. "It seems to me . . . that your solution does not correspond to a physical possibility," he wrote to de Sitter. "In my opinion, it would be unsatisfactory if a world without matter were possible."[3]

Einstein tried to find a flaw in de Sitter's solution, but he found none. He then began to have misgivings about the cosmological constant. If it gave such a strange universe, was it valid? Einstein and de Sitter corresponded frequently over the next year or so. Einstein didn't mince any words about his dislike for de Sitter's model, and in return de Sitter criticized several aspects of Einstein's model. But it was a friendly exchange, and the two men remained close for many years.

Over the next few years the debate among cosmologists was: which model is correct, Einstein's or de Sitter's? Strangely, it was soon shown that Einstein's model was also unstable. It was like a coin balanced on its edge; a slight push in one direction or the other and it would expand or contract.

FRIEDMANN'S UNIVERSE

While scientists discussed the merits of Einstein's and de Sitter's models of the universe, another solution came to light. Aleksandr Friedmann of St. Petersburg, Russia, a meteorologist with a strong background in mathematics, began working on Einstein's theory in the early 1920s. Einstein had added the cosmological constant to stabilize his universe, but he didn't solve the cosmological equations without the cosmological constant. Friedmann did, and he obtained some interesting results. He found that there were three possible fates for the universe, depending on its average density (figure 14). If it was over a certain critical density, the universe was positively curved and would collapse back on itself. In this

Fig. 13: Aleksandr Friedmann.

case it would have spherical geometry and would be closed. If it had less than this critical density, the universe would go on expanding forever. It would be curved like the surface of a saddle (negatively curved) and would be open. If its density was exactly equal to the critical density, it would be flat, but open, and therefore would still expand forever.

Friedmann wrote up his solution in 1922 and sent it to Einstein. At this time Einstein was still convinced that the universe was static, and he

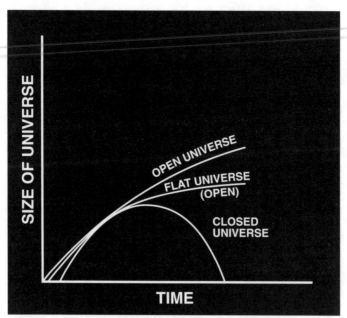

Fig. 14: The three possibilities in Friedmann's model
of the universe: open, flat, and closed.

looked at the new theory as nothing more than an exercise in mathematics. As far as he was concerned, it had little to do with the real universe. Nevertheless, Einstein sent the paper in for publication. Surprisingly, however, after it was published he found what he thought was an
error, and he sent a note to the editor pointing it out. The note was published, and Friedmann saw it; looking carefully through his theory, he
soon saw that Einstein was wrong. With some trepidation he wrote to Einstein, pointing out the problem. Slightly embarrassed, Einstein admitted
in print that he had made a mistake.

There was little interest in Friedmann's model when he published it, but
as we will see, it is the one we use today. It was modified slightly by Howard
Robertson and Arthur Walker of the United States, but the basis of the model
remained unchanged. Friedmann, unfortunately, did not live to see the
importance of his solution; he died in 1925 at the age of thirty-seven.

Another way of looking at the model is to introduce what is called
omega, which is the ratio of the known density of the universe to the critical density (the density corresponding to a flat universe). If omega is

greater than one, the universe is positively curved and closed; if it is less than one, the universe is negatively curved and open. And if it is exactly one, it is flat and open. The term omega is used extensively in modern cosmology. Indeed, one of the major problems of modern cosmology is to determine omega as accurately as possible.

LEMAÎTRE'S UNIVERSE

Only a few years after Friedmann put forward his model, a Jesuit priest in Belgium, Georges Lemaître, came up with another model. Lemaître studied math and physics at Louvain but was also interested in the priesthood, and in 1923 he was ordained. He then left Belgium to study under Arthur Eddington at Cambridge and in the United States at MIT, where he finally obtained his doctorate in 1927. By the time he returned to Belgium he had developed an interest in general relativity and cosmology and was soon working on his own cosmology. Unlike Friedmann, he retained the cosmological constant in his theory. He saw that objects in de Sitter's universe expanded away from one another. In fact, he noticed that according to the theory there was a linear relationship between this velocity and the distance of the object from Earth. He finally arrived at a model that was between Einstein's and de Sitter's models. It began in a state of expansion, then became static as an Einstein model; finally, it began to expand again as a de Sitter model. Lemaître showed his model to both Einstein and de Sitter, but neither showed any interest. He published his paper in 1927, but it was published in an obscure journal and hardly anyone noticed it.

Even as late as 1930 few people had heard of either Friedmann's or Lemaître's models. Most cosmologists were still concerned with Einstein's and de Sitter's models. But there were problems with both of them. Eddington made a comment about the two theories that was published in the journal *Observatory*. It came to the attention of Lemaître, and he wrote to Eddington reminding him about his theory, which he had shown him several years earlier. Somewhat embarrassed, Eddington wrote a note to the journal *Nature* bringing Lemaître's paper to the attention of the scientific world. He later arranged to have it translated and published in *Monthly Notices*.

Within a short time Lemaître began thinking about the beginning of the universe. How did it come into being? He finally decided its initial state had to have been an unstable sphere of neutrons that he referred to as the "primeval nucleus." It was known that when a neutron is not in an atom, it is unstable and decays in about twelve minutes to a proton and an electron. Lemaître spent considerable time studying the idea. He eventually became convinced that cosmic rays, which had been discovered in space, were given off by the primeval nucleus.

SLIPHER AND HUBBLE

While the various theories and models of the universe were being debated, astronomers were learning more and more about the universe. One of the major tools they were using was spectroscopy. Years earlier it had been discovered that when the light of a star or other luminous object in space was passed through an instrument called a spectroscope (the heart of the spectroscope at that time was a prism), lines were seen, and these lines gave valuable information about the object. They told astronomers what elements were in the object, and they also gave them considerable other information. For example, we know that a line from a given element is usually in a particular position, but if the object is moving away from us, or toward us, it will be displaced one way or the other. If the object is moving away from us, for example, the line will be shifted toward the red end of the spectrum; this is called a redshift, and the greater the speed, the greater the shift. Similarly, if it is shifted in the opposite direction (toward the blue end of the spectrum), the object is approaching us.

Percival Lowell, a wealthy Boston businessman, was interested in using spectra to get information about the planets. He became intrigued with Mars after Giovanni Schiaparelli of Italy discovered what appeared to be canals on its surface. Lowell was sure that Mars was inhabited, and he wanted to prove it. After setting up an observatory at Flagstaff, Arizona, under some of the clearest skies in the United States, he hired a young astronomer named Vesto Slipher. He instructed Slipher to obtain the spectrum of Mars and other planets and see what could be determined from them.

Fig. 15: A spiral galaxy. (Photo courtesy of the
National Optical Astronomy Observatories.)

The task was not easy. It took Slipher many months to get everything working properly, but, finally, using the twenty-four-inch refracting telescope at the observatory, he was able to get the spectra of both Mars and

Venus. He was, in fact, able to show that Venus was rotating much slower than previously thought. Lowell was also interested in the "white smudges" in the sky. Some people thought they were "island universes" of stars, so far away that we couldn't distinguish individual stars, but they had no proof. Lowell had other ideas. He was convinced that they were the beginning stages of planetary systems, in other words, systems like our solar system.

One of the most prominent of these "white nebulae," as they were called then, was in the constellation Andromeda, and, as you might expect, it was called the Andromeda Nebula. Lowell instructed Slipher to obtain its spectrum. It was a difficult task, but after considerable work Slipher finally succeeded. To his surprise, he found a tremendous shift in its spectral lines. Most objects at that time showed only small shifts, corresponding to velocities away from, or toward, us (called radial velocities) of about 10 km/sec. The Andromeda Nebula had a redshift corresponding to a velocity of 300 km/sec. The shift in this case was toward the blue end of the spectrum, which indicated that the Andromeda Nebula was approaching us. Slipher informed Lowell, who was now back at Boston, and Lowell encouraged him to continue looking at other nebulae.

Slipher then looked at one in the constellation Virgo and discovered that it had an even greater shift. Furthermore, this time the shift was toward the red end of the spectrum, indicating that the nebula was receding from us. Slipher continued his work, and by 1914 he had the spectra of fourteen nebulae, and almost all were redshifted. He announced his results at the Evanston, Illinois, meeting of the American Astronomical Society. No one was sure of the meaning of his discovery, but most of those in the audience were convinced that it was an important development. He received a standing ovation.

What was Slipher's interpretation of his discovery? He was not sure what the objects were, but he was aware that many astronomers thought they were distant systems of stars, and he eventually began to share their opinion. He was also confused about the fact that a few showed blueshifts, indicating movement toward us, but the majority showed redshifts, indicating movement away from us. He thought this was likely due to the movement of our galaxy, the Milky Way.

By 1922 Slipher had determined the radial velocities of forty-one galaxies, but he was now approaching the limit of his twenty-four-inch

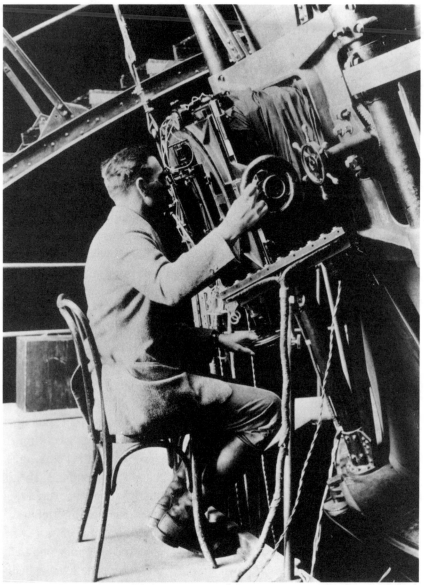

Fig. 16: Edwin Hubble. (Photo courtesy of the
Henry E. Huntington Library and Art Gallery.)

telescope. He did little in the way of speculating on his results, but both Einstein's and de Sitter's theories of the universe were now well known, and it was known that de Sitter's model predicted a recession of distant objects and redshifts. Furthermore, Slipher's data seemed to indicate a crude relationship: the dimmer the galaxy, the greater the redshift. But at this stage there were still too many unknowns. Astronomers still didn't know for sure what the nebulae were or whether they were inside or outside our system, the Milky Way.

When Slipher first announced his results in Evanston in 1914, a young graduate student was in the audience. His name was Edwin Hubble, and he had been doing his thesis on white nebulae. He graduated soon after the meeting and was planning on continuing his study of nebulae, but World War I intervened. After serving in the war, however, he returned to Mt. Wilson in California. The giant one-hundred-inch Hooker telescope had just been completed, and Hubble soon had access to it, along with the sixty-inch reflector at the summit of Mt. Wilson. He was determined to find out what the white nebulae were. Were they "island universes" of stars, as some people believed, or were they just large regions of gas, or were they the early stages of a planetary system?

Hubble made good use of the two huge telescopes on the mountain. He took longer and longer exposures of the nebulae, beginning with the one in Andromeda. Some of them began to show what looked like stars, but Hubble needed more proof. The evidence was still weak. He thought he would look for "novae," or exploding stars, since they were much brighter than ordinary stars and would easily show up on long exposures. Furthermore, they would be strong evidence that there were, indeed, stars in the nebula. As he compared plate after plate, he soon discovered that some of the starlike objects in the nebulae were changing in brightness, over a regular, or constant, period of time. They were *variable stars.*

Variable stars had been studied extensively over the years, and their properties were fairly well known. In particular, a type called Cepheids had been discovered, and it had been shown that they satisfied a period-luminosity relationship. In other words, their period was related to their absolute, or true, brightness. This meant that all you had to do was measure the period and you could determine the distance to the Cepheid. You would therefore know the distance to the nebula that it was in. Although the relationship was crude at this stage, it was a valuable tool.

Hubble soon convinced himself that many of the variables he was seeing in Andromeda and other galaxies were Cepheids, and with a little work he was able to determine their distance. Using twelve Cepheids in the Andromeda Nebula, he determined its distance was 900,000 light-years (we now believe it is about 2 million light-years away). This was considerably larger than our galaxy and meant that the Andromeda Nebula had to be a distant system, well outside ours. Hubble went on to study other nebulae. They were even farther away. Soon it was obvious that the white nebulae were, indeed, systems of stars, each well beyond the limits of our system. The universe appeared to be populated by millions, perhaps billions, of these systems. We now refer to them as *galaxies*.

Hubble then turned to the radial velocities of these galaxies; he knew that Slipher had obtained a large number, and he took advantage of them. Using Cepheids, Hubble was able to determine the distance to six of Slipher's galaxies, and to a first approximation, it appeared that the greater the redshift, the more distant the galaxy. But was this true universally? Hubble wanted more proof, and for this he knew he would have to probe farther out into the universe. In these more distant galaxies Cepheids were not visible, so Hubble turned to the brightest stars in them. He knew that to a first approximation these stars would all be about the same brightness. Using this, he was able to determine the distance to another fourteen galaxies. He then made a plot of distance versus velocity of recession. Although there was a lot of scatter in the points, they were generally in a straight line. The slope of the line is now called the Hubble constant and is denoted by H. At this stage things were still quite crude; nevertheless, it seemed that all galaxies were moving away from us, and the farther they were away, the faster they were traveling.

But what about the Andromeda galaxy? It, and a few others, showed blueshifts. As it turned out, galaxies frequently form larger structures known as clusters of galaxies. The galaxies within a cluster are held together by gravity. The Andromeda galaxy and a few others are in our "Local Cluster." Because the galaxies within the cluster are held together gravitationally, they are not expanding away from one another.

Hubble published his results in 1929. De Sitter's prediction of an expanding universe was well known at the time, and Hubble referred to it at the end of his paper.[4] But he knew he still had a lot of work to do. He took on an assistant by the name of Milton Humason, and together they

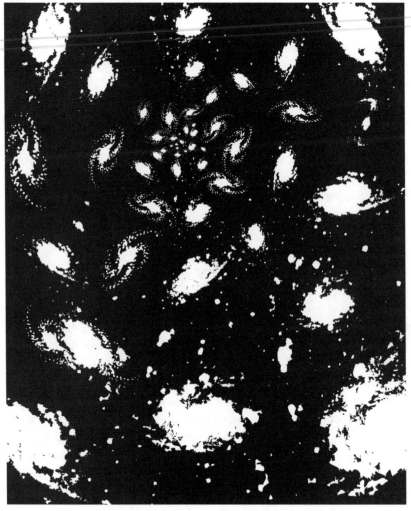

Fig. 17: The expanding universe with galaxies moving away from one another.

probed even farther into space. By 1931 they had obtained thirty-seven more radial velocities and distances, and Hubble made a new plot. This one was much more accurate than the previous one, but it still had some scatter in the points. Nevertheless, to a good approximation, the relationship was still linear (a straight line could be drawn through the points). The line now reached out to 100 million light-years.

There was no doubt that the universe was expanding. It was not static as Einstein and de Sitter had assumed. The evidence was soon overwhelming, but Hubble was cautious. By 1936, however, when he published his book *Realm of the Nebulae*, the idea of an expanding universe had gained almost universal acceptance by astronomers.[5]

When Einstein heard about the expansion, he realized that the introduction of a cosmological constant had been a mistake. It wasn't needed. Indeed, if he hadn't introduced it, he may have been able to predict the expansion of the universe. He later referred to it as the "greatest blunder of his life."[6]

THE BIG BANG

Lemaître was the first to put forward the idea that the universe began as a gigantic explosion from a nucleus. His nucleus, which he referred to as the "primeval nucleus," was composed of neutrons and was about the size of the solar system. Others eventually began to think of it as much smaller than that, until, finally, many astronomers began to suggest that it began as a tiny point—a singularity. It was hard for many people, however, to accept that the universe had arisen from a singularity.

Among those pondering the beginning of the universe was George Gamow. He had recently immigrated to the United States from Russia and was an expert on nuclear physics. He developed a view quite different from Lemaître's; he became convinced that the elements of the universe were built up in collision processes in the early universe. His "initial state" was a mixture of protons, electrons, and neutrons, along with radiation that he referred to as *ylem* (which means "first substance"). It was obvious to him that temperatures in the early universe would have been exceedingly high but would have dropped rapidly as the universe expanded. According to his calculations, the temperature would have been about a billion degrees when the universe was five minutes old. This was much too high for colliding particles to stick together. But as the expansion continued and the temperature lowered, protons and neutrons eventually would stick together producing deuterium. Then tritium would be created, and then helium, as other particles struck the tritium nucleus.

Gamow was soon convinced that all of the elements of the universe

Fig. 18: George Gamow.

were produced in this way. He wrote a paper outlining his ideas in 1946. Then, over the next couple of years, he developed the theory further with graduate student Ralph Alpher. As a result of a pun, the theory eventually became known as the alpha-beta-gamma theory, after the first three letters of the Greek alphabet. Gamow, who was a great practical joker, noticed that his name sounded like the Greek letter gamma, and Alpher's sounded like alpha, but he needed a beta to complete the sequence. As it turned out, he was friends with Hans Bethe of Cornell University, so he put his name on the paper, even though Bethe had nothing to do with it.

Gamow had barely finished the paper, however, when Enrico Fermi

Fig. 19: A joke Gamow's students played on him. Ylem is the material out of which the universe was formed. (Photo courtesy of Ralph Alpher.)

and one of his graduate students noticed a problem. There were "gaps" at atomic numbers 5 and 8 that couldn't be overcome. In short, new elements would not form at these points because particles wouldn't "stick" when they struck these nuclei. Gamow realized the problem about the same time. It appeared, therefore, that a few of the light elements such as deuterium, tritium, and helium could be produced in the early universe, but nothing else.

Where and how, then, were the other elements formed? The problem wasn't solved for another nine years. In 1957 Fred Hoyle, Geoffrey and Margaret Burbidge, and William Fowler showed that they could be produced in massive stars. Furthermore, such stars eventually exploded as supernovae, distributing them to space, where they could form solar systems such as ours.

Evidence for the big bang theory, as it was soon called (it was so named by Hoyle in 1948), seemed to be overwhelming. But for a while it had a rival. Fred Hoyle, Hermann Bondi, and Thomas Gold of Cambridge University put forward what they called the steady state theory. It assumed that the universe has always been the same; in other words, it is in a steady state. For the universe to remain in a steady state, however, matter that is lost as the universe expands had to be replaced. Hoyle, Gold, and Bondi proposed that small amounts were introduced in the vast regions between the galaxies. Only a small amount was needed. As we will see, however, the big bang theory eventually triumphed over the steady state theory.

The big bang theory has, indeed, explained many things about our universe, and it has had many successes. One of its greatest predictions was what is called the cosmic background radiation. Gamow and Alpher first predicted it; according to their calculations, it was given off early on and cooled as the universe expanded. It would now have a temperature of only a few degrees above the lowest possible temperature, absolute zero. Gamow was told by astronomers that there was little chance of detecting it, so he didn't look.

A few years later, in the early 1960s, Robert Dicke of Princeton University was studying the early universe when he also realized that "fireball" radiation would be given off in the initial explosion, and he soon became convinced that it should be detectable. He asked colleague Jim Peebles to determine the temperature accurately, and Peebles came up with about 3K. He then asked two experimentalists at Princeton, Peter Roll and David Wilkinson, to search for it. They had barely begun, however, when Dicke found out it had already been discovered.

Arno Penzias and Robert Wilson of Bell Labs had discovered the

Fig. 20: Penzias and Wilson in front of the Holmdel telescope.
(Photo courtesy of Bell Labs.)

radiation using a strange, hornlike radio telescope at Holmdel, New Jersey. When they first detected it, the radiation was nothing more than a nuisance, as far as they were concerned; it was a strange "hiss" in their telescope. They were annoyed because they were having difficulty getting rid of it. Dicke heard about their problem through a mutual friend and visited the telescope, and soon there was no doubt: the cosmic background radiation that Dicke and Peebles, and, earlier, Gamow and Alpher had predicted, had been discovered. It had a temperature of approximately 3K and was soon shown to be slightly anisotropic (different in different directions). The satellite COBE (Cosmic Background Explorer) verified the discovery several years later and gave the complete curve for its intensity at all wavelengths.[7]

The discovery of the cosmic background radiation was a tremendous success for the big bang theory; not only did it predict the radiation, but it correctly predicted its temperature. Furthermore, the discovery ruled out the steady state theory, which did not predict it. Despite the success, astronomers soon realized that the big bang theory had other problems.

The first, called the horizon problem, was noticed by Charles Misner in 1969. He determined that the two opposite edges of the universe were not causally connected because light could not have traveled from one edge to the other during the time the universe has existed. This meant that they could never have been in communication with one another, yet the cosmic background radiation at both locations had the same temperature. How was this possible?

A second problem, now called the flatness problem, was discovered by Robert Dicke in 1978. To understand it, let's go back to the quantity omega that we introduced earlier; it is the average density of matter in the universe divided by the critical density. As we saw, if omega is greater than one, the universe is closed; if less, it is open. When it is exactly equal to one, the universe is flat and open. Dicke showed that if the universe is flat now, it had to have started out being *exactly* flat, with omega being 1.0000 to about fifty zeros. If it started out slightly above one, it would be huge now—of the order of millions. On the other hand, if it started out slightly less than one, it would now be tiny—approximately zero.

These are the two major problems of the big bang, but there are others. A third problem is: how did the galaxies form? We now believe they were created by tiny fluctuations in the original cloud of gas and particles. But the big bang theory gives us few details.

INFLATION THEORY

For several years scientists were stumped by these problems. Then a breakthrough came in late 1979 with the formulation of what is now called inflation theory. The basic insight was due to Alan Guth, now at MIT. In 1978 Guth attended a lecture in which Dicke discussed the flatness problem. The idea impressed him, but he was skeptical of cosmology at the time; to him it was too speculative. He was interested in particle physics, which was on a much firmer footing than cosmology, as far as he was concerned. Shortly after he heard Dicke's lecture, however, he began working on a problem with a colleague, Henry Tye, and it soon took him into the arena of cosmology. The two men became interested in how magnetic monopoles (heavy particles with single magnetic poles) were generated in the early universe. Little was known at the time, and because of this, they decided to take a closer look at the accepted model of the early universe.

It soon occurred to them that a "supercooling" had to have occurred in the early universe. Supercooling occurs in liquids, including water, under certain circumstances. Water usually freezes at thirty-two degrees Fahrenheit, but it can be supercooled to several degrees below thirty-two before it freezes. If supercooling occurred in the early universe, they realized that it had to be in a state referred to as a "false vacuum." In December 1979 Tye left the country, and Guth was on his own. Late one night as he was filling his notebook with equations, he realized that if the universe did enter a false vacuum, a negative pressure would have been created, which in turn would have generated a negative gravity. The result would have been a sudden and dramatic expansion of the universe. For a brief period of time after 10^{-35} second it would have expanded at an incredible rate—much faster than it had been expanding. But this "inflation," as it was called, would be over by 10^{-33} second.

What delighted Guth was his finding that inflation solved the major

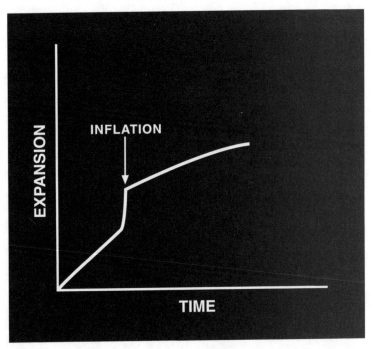

Fig. 21: A simple representation of inflation in the early universe.

difficulties of the big bang theory. It solved both the flatness and the horizon problems, and it also helped solve the problem he had been working on related to magnetic monopoles. Furthermore, it gave the first hint of how the universe may have begun—in particular, where its initial energy came from. Guth showed that the energy was self-generating.

Since the original inflation theory, many variations have been proposed. Most of them assume that numerous "bubbles" were generated during inflation, and our universe is one of these bubbles. This implies that there are many universes out there, besides ours. There is no way, however, that we will ever be able to contact anyone in any of them.

One of the most troublesome predictions to come out of inflation theory is its prediction that the universe is flat; in other words, omega is one. As we will see, this has created many problems and difficulties over the past few years.

DARK MATTER

The major difficulty caused by the prediction of a flat universe is that we know that all the visible matter in the universe adds up to less than 1 percent of that needed to make the universe flat. For years, however, we have known that there is a lot of matter out in the universe that we can't see directly. The Dutch astronomer Jan Oort first noticed a problem in 1930 when he tried to estimate the mass of our galaxy by determining the gravitational pull it exerted on stars that were straying outside the disk of the galaxy. He determined that we were seeing only about 50 percent of the mass of our galaxy. In 1931 Fritz Zwicky of Mt. Wilson Observatory was observing clusters of galaxies when he realized that they had to have several hundred times their observed mass to remain bound. Then in 1972 Jeremiah Ostriker and Jim Peebles of Princeton University showed that spiral galaxies, such as our Milky Way, would be unstable unless they were surrounded by an invisible halo of matter. And finally, about the same time, Vera Rubin and Kent Ford of the Carnegie Institution of Washington found evidence for a huge halo of what we now call "dark matter" around several nearby galaxies.

The evidence is now overwhelming. The universe is composed of a lot of dark matter—matter that cannot be seen directly but signals its presence through its gravitational field. This has created a serious problem for the big bang theory. If the universe is flat, as predicted by inflation theory, 99 percent of its matter has to be dark. Is there enough dark matter out there to account for this? What form does the dark matter take? Astronomers soon narrowed their search to several candidates: small brown dwarf stars, planetlike objects, black holes, and various types of particles. But in the 1970s nucleosynthesis arguments (based on the abundance of light elements in the universe) showed that ordinary matter—made up of protons and neutrons—could account for only about 10 percent of the critical density. Most of the dark matter, assuming it existed, had to be in a strange, exotic form. Soon several types of particles were invented to account for it, but many astronomers were skeptical.[8] Then came an even greater difficulty.

THE ACCELERATING UNIVERSE

In the late 1990s, much to the surprise of astronomers, observations began to show that the expansion of the universe was accelerating, rather than decelerating as expected. The main evidence for this came from a study of what are called Type Ia supernovae. Supernovae are stars that explode, and Type Ia supernovae are explosions that occur in relatively small stars about the size of our sun. Most supernovae occur in stars that are hundreds of times more massive.

All stars are known to collapse on themselves when they begin to run out of fuel. This will eventually happen to our sun, but it won't happen for millions of years, so you don't have to worry. When a star like our sun collapses, it eventually becomes a white dwarf, about the size of Earth, or perhaps a little larger. Most white dwarfs fade away gradually over millions of years, disappearing as cosmic ash. But like people, stars like to congregate; indeed, most are in double or triple systems, or even systems of many stars. And if a white dwarf is in such a system, it can capture some of the escaping gas from its companion, and if this happens, a gigantic explosion may occur. In short, the star becomes a Type Ia supernova. Supernovae of this type are not all of the same absolute brightness, but we know that the brightest ones last longer, so we can adjust our calculations. Because of this, Type Ia supernovae are now one of our best standard candles. (Standard candles are luminous objects that are used to determine the distance to galaxies, and so on, because they are of known magnitude.)

Several studies have been made using Type Ia supernovae. Both the Keck telescope and the Hubble space telescope have been used. Groups headed by Saul Perlmutter of Lawrence Berkeley National Laboratory in California, Neta Bahcall of Princeton University, and Ruth Dolby also of Princeton University have made studies, and the evidence is now clear. In all cases, these studies indicate that the expansion of the universe is accelerating. This is confusing, to say the least; if the universe were composed mainly of matter, or ordinary energy, its expansion would be slowing down, or decelerating, because of the mutual gravitational pull of the matter.

It almost seems impossible that the universe could be accelerating. What is causing it? One answer comes from Einstein's cosmological theory; in it, as we saw, Einstein introduced the cosmological constant. He used it to stop his model of the universe from collapsing (or

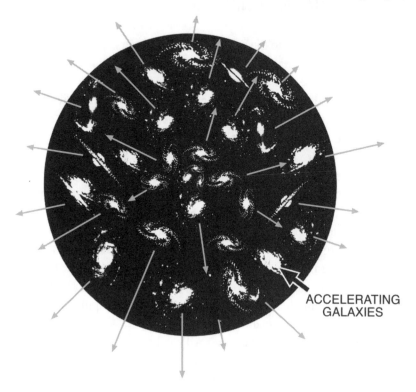

ACCELERATING GALAXIES

Fig. 22: Acceleration of the outer galaxies in the universe.

expanding); it acted like antigravity in that it exerted an outward force opposite to gravity. After it was discovered that the universe was expanding, Einstein rejected the constant, but others didn't agree with him; they preferred to keep it. One of the reasons was the discovery of quantum electrodynamics, which deals with the interactions of particles and radiation. In formulating this theory, scientists noticed that "empty space" wasn't as empty as they had thought; it was actually alive with what are called "virtual particles." These particles appear and disappear too rapidly to be seen, but we know that they *have* to exist. They are essential for particle reactions, which we can see. And of particular importance, they represented a tremendous amount of energy. This energy is represented in Einstein's equation by the cosmological constant.

Astronomers prefer to refer to this energy as "dark energy."[9] At one time scientists thought that most of the universe was made up of dark matter. They were surprised when deuterium measurements showed that, even with ordinary dark matter, the universe was far short of the amount needed to close it. Most of the dark matter had to be in an exotic form. Now it seems that things are even more complex. If the universe is flat, as predicted by inflation theory, most of the universe cannot be in the form of matter of any type; most of it has to be energy, and since we know little about the form of this energy, astronomers sometimes call it "funny energy."

There is, however, a problem. If the universe is presently accelerating, and this acceleration is due to the cosmological constant, we have to ask about the sudden acceleration that occurred at the beginning of the universe that we refer to as inflation. Was it also due to the cosmological constant? If so, the cosmological constant would need two different values in the two eras. It couldn't be a constant, but has to be variable, and most scientists do not like this possibility.

In short, according to recent estimates, slightly over 70 percent of the universe consists of dark energy, and we have no idea how it arises or exactly what it is. Of the remainder, about 27 percent is dark matter, and most of it is in the form of exotic particles that we know little about. Finally, less than 1 percent is ordinary matter and radiation. Clearly, we have a serious problem.

QUINTESSENCE

Because problems persisted with vacuum energy, or equivalently, the cosmological constant, several scientists, including Robert Caldwell, Rahul Dave, and Paul Steinhardt of the University of Pennsylvania, invented a new field. They call it *quintessence* after a fifth element that was hypothesized by ancient philosophers. Quintessence obviously has to be similar to vacuum energy, but it differs in that it can change or evolve. Furthermore, it differs in several other ways, and it is these differences that will allow us to determine if the acceleration is arising from quintessence or from vacuum energy. First of all, as we noted, quintessence can evolve; the changes are likely to be extremely small, and to a first approximation, it would look constant to us. Vacuum energy does not evolve. Second, waves can propagate through quintessence; they cannot propagate through vacuum energy. Also calculations show that vacuum energy would cause a greater acceleration rate than quintessence. Thus by measuring the acceleration rate accurately, we should be able to determine what is causing it. Supernovae and the cosmic background radiation will also be helpful in this respect.

Does the fate of the universe depend on which of the two is accelerating the universe? Indeed, it does. If it is vacuum energy, our present ideas are valid, and there is nothing new to learn. If, on the other hand, it is quintessence, the end could be quite different. We're still not sure what happens to the universe if quintessence prevails.

ALTERNATE THEORIES

As you might expect, there are several alternatives to inflation and its variations. Although these theories are not as developed as inflation theory, they do overcome many of the same problems. João Magueijo of Imperial College, London, has put forward a variable light speed theory.[10] It's well known that most of the so-called constants of nature have, at one time or another, been assumed to vary. Paul Dirac put forward a theory in the 1930s hypothesizing that the gravitational constant varied, and there have been theories that assumed varying electronic charge and mass.

Strangely, though, until recently few have suggested that the speed of light might vary. The main reason is that it is at the foundation of relativity theory, and few wanted to tamper with this. (Interestingly, though, as we saw earlier, Einstein's first generalization of his special theory of relativity was a variable light speed theory.)

Magueijo has shown that if the speed of light was slightly higher in the very early universe, both the horizon and flatness problems are solved. No inflation is needed; the universe could expand uniformly. Furthermore, the change in the speed of light would be small. We, of course, have no evidence for a slightly greater speed of light at this time, but we also have no evidence for an inflation at this time, either.

How can we find out if this idea is correct? Again, only experiment can tell us, and so far the predictions of the variable light speed theory are uncertain because the theory needs further development. According to many scientists, however, it is an idea worth pursuing.

Another somewhat radical idea was put forward by Mordehai Milgrom.[11] It is directed at dark matter, rather than dark energy, but it could have serious consequences on cosmology in general if found to be correct. As is well known, the dark matter problem arose because of the gravitational fields of galaxies and clusters of galaxies. According to the mass they appear to have, their gravitational pull is not sufficient to stabilize them. The effect is seen throughout the universe, from dwarf galaxies all the way up to superclusters of galaxies, and the larger the mass, the greater the discrepancy. Adding to the confusion, researchers have determined that most of the dark matter has to be in an unknown, exotic form. Ordinary matter—protons, neutrons, and so on—won't do.

Milgrom has shown that we can overcome the dark matter problem if we make a slight adjustment in Newton's law of gravity and his second law of motion. According to Newton's law of gravity, masses attract one another as the square of the inverse of the distance between them. Milgrom has shown that a slight adjustment in this law would overcome the problem. Indeed, the change is so slight that we would have trouble measuring it. According to his theory, if the acceleration is greater than a certain value a_0, Newton's law is valid, but if it is less, the law has to be slightly changed. The constant a_0 is extremely small.

This is, of course, not the first time that an adjustment has been made to Newton's laws, so it's not a particularly large step. Einstein made an adjustment when he formulated relativity theory, and changes were made in the microworld when quantum mechanics was formulated.

Most scientists don't take these alternatives too seriously, but they do give us food for thought.

WHAT WOULD EINSTEIN THINK?

Much has changed in cosmology since Einstein's theory of 1916, but *all* cosmologies are still based on his theory of general relativity. Little was known about the universe when Einstein formulated his theory, so he had to make some rather daring hypotheses. Some of his ideas are now known to be wrong; he didn't know that the universe was expanding and assumed it to be static. But his hypotheses of homogeneity and isotropy are still at the foundations of all cosmologies. We have learned a tremendous amount about the universe since his time, but mysteries and puzzles still abound. We still don't know what most of the universe is composed of. We're still not sure whether the universe is open or closed, or whether it is flat as predicted by inflation.

It's always interesting to consider what Einstein would have thought about the present situation. No one can know for sure, but we do know he liked simplicity and beauty in his theories, and he was, without a doubt, a daring innovator. He was one of the first to accept the idea of the quantum, when almost no one took it seriously. He showed us that time and space were completely different than we had assumed. He introduced the idea of curved space, even though the idea was laughed at when he first introduced it.

Would he be attracted to one of the alternate theories, which seem simpler? We cannot be sure, but we do know that it will take somebody with Einstein's insight to see through the present difficulties.

Chapter 3

Black Holes,
Wormholes, and
Other Demons

A s we saw earlier, Schwarzschild obtained two solutions to Einstein's equations. One was for the exterior region of a star, and the other for the interior. He noticed immediately that there was a singularity in the solution, in other words, a region where things went to infinity and the theory broke down. Singularities were not new; Newton's gravitational theory had a singularity in it, and so did Maxwell's electromagnetic theory. But they were singularities at the center of the mass or charge. Schwarzschild found a singularity at the center, but he also found one that was at a finite radius, and this meant that the region inside the singularity was cut off from the outside world. This was something new to science.

Schwarzschild was disturbed by the singularity; he even went as far as calculating the singularity's radius for the Sun, getting a value of about three kilometers, or equivalently, two miles. (This radius is now referred to as the gravitational radius.) In theory, if all the matter of the Sun could be compressed inside three kilometers, we would not be able to see it; it would be cut off from us. Schwarzschild went on to consider a static distribution of matter of uniform density, determining that the pressure at the center would become infinite before the gravitational radius was reached if it was compressed. To him this meant that the gravitational radius was inaccessible and could therefore be ignored.

Einstein studied Schwarzschild's solution, and, as you might expect, the singularity also bothered him. Why was there a region in the center of a spherical mass that was cut off from us? It didn't make sense. But Einstein said very little about the region for several years. It appears that he was worried it might make the solution invalid and might be a serious problem for the theory.

About a year after Schwarzschild discovered his solution, a student, Johannes Droste, who was working for the physicist H. A. Lorentz in Holland, independently obtained the same solution, but he went further than Schwarzschild. He examined the trajectories of particles and light rays in the curved space near the gravitational radius and found that at 1.5 times the gravitational radius light rays would take up circular orbits around the object. This meant that if these rays were coming from the object, they could not escape. He also found that, as seen by an outside observer, particles falling toward the object would never reach the gravitational radius (as seen by an outside observer). The "hole" in the center was therefore inaccessible, and like Schwarzschild, he assumed that if it was inaccessible it could be ignored.

In the same year, Ludwig Flamm of Germany took a look at the geometry of the curved space around the singularity. He discovered that it looked like a funnel that ended on the gravitational radius. Shortly thereafter, Hermann Weyl of Zurich went even further. He showed that the funnel was two-sided; in other words, there was a funnel on both sides of the circle.

Over the next few years several people, including David Hilbert of Göttingen, worked on the problem, but little progress was made. No one was quite sure what this strange region was. It was soon referred to as the "forbidden region." Then Cornelius Lancos, who was Einstein's assistant for a while, added to the confusion. He applied a transformation of coordinates and found that the singularity shifted position. Was it a true singularity? True singularities didn't do this.

Einstein heard of these developments, but he continued to say little. Finally, in 1922, during a conference in Paris, the French mathematician

Jacques Hadamard cornered him and asked him what he thought. Einstein had to have been perplexed. He replied, "If that term could actually vanish somewhere in the universe, it would be a true disaster for the theory."[1] (He later referred to the problem as the "Hadamard disaster.") The problem bothered him so much that he made some calculations. He was sure the region inside the gravitational radius was inaccessible and could be ignored, and sure enough, his calculations showed him that it was impossible for a physical system to lie completely inside the gravitational radius. He announced his results at the meeting on the following day.

Einstein's opinion was, of course, extremely important, since he had created the theory. But Eddington was also a well-known expert in the area, and his opinion was also valued. At first he was quite whimsical about the problem. In his book *The Internal Constitution of Stars*, which was published in 1926, he discussed it in considerable detail.[2] He acknowledged that because of the strong gravitational field, light rays would not be able to leave such an object, and we would therefore not be able to see it. He went on to say that the curvature of space would be so great, it would actually "close up around the star." We know that space does not wrap itself around the star, and Eddington was no doubt just trying to be a little coy. In the end, though, Eddington was as strongly against the existence of such objects as Einstein was. He was sure they couldn't exist, but he was, nevertheless, intrigued with the region and referred to it as the "magic circle."

An important breakthrough came a few years later. Lemaître became interested in the problem, and in 1932 he showed that the Schwarzschild singularity wasn't a true singularity after all. Its existence depended on the coordinate system that was being used. In other words, it was a singularity in some systems, but not others. As in the case of his cosmology, however, he published the result in an obscure journal, and nobody noticed it for several years. H. P. Robertson of Harvard finally took note of it and brought it to a wider audience. But if it wasn't a true singularity, the region inside the gravitational radius was indeed accessible, and this brought renewed worries.

Einstein decided to take another look at the problem in 1935. Working with Nathan Rosen, he decided to study the geometry of the space around the gravitational radius. As Droste had earlier, he found a funnel-shaped tunnel that led to the gravitational radius. To his surprise,

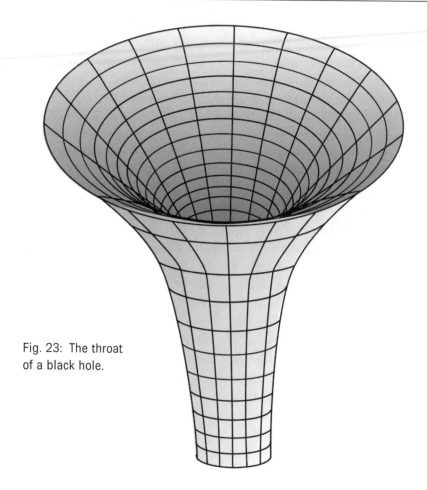

Fig. 23: The throat
of a black hole.

however, he discovered that there was a "mirror image" tunnel on the
other end of the gravitational radius. The question immediately arose:
where does the tunnel lead? To Einstein the only answer seemed to be "to
another universe." But this didn't make sense to him, and he didn't like
it. He therefore determined the velocity that would be needed to pass
through the tunnel, and it turned out to be greater than the velocity of
light. Einstein was relieved; it meant that the region was inaccessible and
could be ignored. These tunnels soon became known as "Einstein-Rosen
bridges." Later they were called *wormholes* in space.

Still, the problem didn't go away. After Robertson heard about
Lemaître's discovery that the singularity was not a true singularity, he

began lecturing on it. In 1939 J. L. Synge of Ireland attended one of Robertson's lectures and heard of the development. He told Einstein of the discovery, and Einstein became alarmed. He was still convinced that the region was inaccessible and again tried to prove it. This time he considered a cluster of stars. Each of the stars in the cluster attracted each of the others gravitationally, and the stars moved in circular orbits around a common center. Einstein assumed the cluster collapsed, and as its radius decreased, the stars moved faster. He showed that at 1.5 times the gravitational radius, the stars would have a speed equal to that of light, and of course nothing could travel faster than that. To reach the gravitational radius, they would have to travel at speeds greater than that of light, and since they couldn't, the region inside this radius was inaccessible.[3]

Einstein was pleased. He had once again proved that this region could be ignored, and as it turned out, he never considered the problem again or even talked about it. Einstein's calculation was, indeed, correct; nevertheless, he made a mistake in judgement. He assumed that there was a force countering gravity, namely, centrifugal force, and he did not consider the possibility of an overwhelming implosion, which we now know occurs. Several years later the objects that Einstein was trying so hard to disprove became known as *black holes*.[4] Einstein was sure they couldn't exist, but his main concern was that their existence would somehow destroy general relativity. General relativity, of course, hasn't been destroyed; moreover, through it we have learned a lot about black holes. Interestingly, the same objects were predicted many years earlier—well before general relativity was published—by John Mitchell of England and Pierre Laplace of France. Both men based their prediction on Newton's theory of gravity, and indeed such objects do exist within Newton's theory, but it tells us little about them.[5]

NEVER-ENDING COLLAPSE

In the same year, 1939, that Einstein was trying to prove that black holes couldn't exist, another physicist, J. Robert Oppenheimer, and his student, Hartland Snyder, were showing that they could. They showed that they would be formed in the final collapse of a giant star.

Oppenheimer was born in New York in 1904. His parents were well-to-do, and he was catered to by servants and maids. It was obvious from an early age that he was extremely intelligent. Always at the top of his class, he rarely got anything but A's. Early on he developed a tremendous appetite for knowledge, but his major interest was science. He collected rocks and minerals and enjoyed the chemistry lab. By the time he graduated from high school, he was fascinated with chemistry, and he decided to major in it at university.[6]

He entered Harvard University in 1922, completing the four-year course in three years. Although his primary interest was still chemistry, during his third year he took a course in physics from Percy Bridgeman and was soon enthralled by both Bridgeman and physics. When he graduated, he therefore decided to go into physics rather than chemistry. He wanted to attend Cambridge University in England, and with the help of an excellent letter of reference from Bridgeman, he was admitted to Ernest Rutherford's lab. He was disappointed, however, when he arrived in England and found that he wouldn't be working for Rutherford. Rutherford had assigned him to another experimentalist, G. P. Thomson.

Oppenheimer had a sharp mind and tremendous mathematical ability, but he was quite clumsy in the lab, and before long he began to hate Thomson's lab and the project he was assigned. He became so depressed that he considered quitting. Then one day Niels Bohr of Denmark visited Cambridge. Oppenheimer talked to him and was thoroughly impressed. Bohr told him about the exciting developments that were taking place in quantum theory. Oppenheimer decided that he was wasting his time doing experimental physics; his real interest was theoretical physics, and the major centers of theoretical physics were on the continent. He packed his bags and left for Göttingen in Germany.

Göttingen was one of the major universities in Germany, and many of the best-known physicists in Europe had trained there. Werner Heisenberg had just formulated his theory of quantum mechanics while working for Max Born of Göttingen. Oppenheimer was also soon working for Born, and within a short time the two men published an important paper in quantum mechanics. Oppenheimer received his doctorate at Göttingen in 1928 and was one of the few people in the world who was an expert in the new branch of physics called quantum mechanics. When ready to return to the United States, he was therefore in great demand and had about a dozen offers. He finally decided on a joint appointment with the

Fig. 24: J. Robert Oppenheimer.

California Institute of Technology and the University of California at Berkeley.

At first, Oppenheimer was not a good teacher; he went too fast for the students and had a tendency to mumble. But he was one of the few experts on quantum mechanics in the United States, and students were soon flocking to him. Gradually his teaching technique improved, and he soon had a dozen or more graduate students working for him. He shared more than physics with them; he took them to cafés where he frequently talked about food, literature, art, and other things. They soon became known as "Oppie's Cronies." His classes became so popular, with some students retaking his quantum mechanics class so often, that he had to literally kick them out of it to make room for others.

Oppenheimer soon had his students doing theses on a wide range of subjects. He would assemble them in his office each day and discuss their progress while other students made comments. Among Oppenheimer's interests was a discovery that had been made a few years earlier by Subramanyan Chandrasekhar. Chandrasekhar had shown that a star with a mass less than 1.4 solar masses (mass of the Sun) would collapse over millions of years to what was called a white dwarf. It would only be about thirty thousand miles across and would have an incredibly high density. The collapse had never been looked at relativistically, however, and Oppenheimer was sure relativity would give greater insights into it. He assigned the problem to George Volkoff, a recent immigrant from Russia who had come to Berkeley from the University of British Columbia in Canada. Volkoff solved the problem within a short time, showing that Chandrasekhar's conclusions were valid even when relativity was applied.

But there was another problem. What about stars more massive than 1.4 solar masses? The Soviet physicist Lev Landau had already speculated on a denser type of star, composed of neutrons. Furthermore, Mt. Wilson astronomer Fritz Zwicky had published a number of papers predicting that such stars should exist. But Zwicky had irritated almost everyone he came in contact with, and few people paid attention to him. When Oppenheimer heard of Landau's work, he knew he would have to continue the problem he had assigned to Volkoff. Landau was a world-renowned physicist, and his opinion was highly respected.

What did relativity say about stars with a mass greater than 1.4 solar masses? Oppenheimer selected one of his brightest students, Hartland Snyder, to work on the problem. Again, the idea was to apply general relativity to see what it predicted. The result was a surprise to both Snyder and Oppenheimer. Interestingly, the year was 1939, and at this very moment Einstein was trying to prove that black holes couldn't exist. But Snyder's result would prove that Einstein was wrong.

To make the problem tractable, Oppenheimer and Snyder made a number of simplifying assumptions. They considered the final collapse of a massive star, which was assumed to be perfectly spherical, of uniform density, and nonspinning. They also assumed that there was no internal pressure and no radiation emitted by the star. The calculation was long and difficult, but Snyder prevailed. Unlike the star that gave rise to a white dwarf, this star did not take millions of years to collapse. It imploded rapidly, in a fraction of a second, but the implosion was strange, to say the least. The collapse was rapid at first, but the closer the star got to its gravitational radius, the slower it collapsed. For someone watching it at a distance, the star would appear to get closer and closer to its gravitational radius, but it would never quite reach it. The reason is that the time on the surface of the star, as seen by a distant observer, was running slower and slower as the star collapsed. Indeed, according to general relativity, it stopped at the gravitational radius, and since no time was passing, it wasn't changing; the star was frozen in time.

Surprisingly, though, when they considered an observer moving with the collapsing star, things were quite different. Time appeared to run normal for him. In a relatively short time, he would pass though the surface of the black hole and into its interior. So the interior was accessible after all! Einstein must have been annoyed when he heard this, but there is no record of his reaction. Most scientists found it strange that time could run at two different rates for two different observers. Oppenheimer was confused about the result, but he never followed up on it. Within a short time, America was involved in the war, and Oppenheimer's attention turned to the building of the atomic bomb. He never returned to the problem after the war.

AFTER THE WAR

Oppenheimer's results were soon forgotten. They were too "exotic" for most people, and many scientists thought they were far-fetched. Although they were interesting, they were hard to take seriously, and as a result, few advances in the area were made through the 1950s. Then in the early 1960s strange signals were discovered in the sky by radio astronomers.[7] They were soon shown to be extremely energetic objects in the outer regions of the universe. Astronomers had no idea how anything could generate so much energy. No known source could produce energy at the rate they were producing it. Then a number of people remembered Oppenheimer's calculations. Perhaps the strange new objects were collapsing stars. Huge amounts of energy would be produced in the collapse of a very massive star.

One of those who became interested in the problem was John Archibald Wheeler. Born in Jacksonville, Florida, in 1911, Wheeler obtained his PhD in 1933 at Johns Hopkins University. After a postdoctoral fellowship with Bohr in Copenhagen, Wheeler joined the faculty of Princeton University in 1938. He was skeptical of the idea of black holes at first, but, as he studied them in more detail, he finally became convinced that they had to exist. He was particularly interested in Oppenheimer and Snyder's paper and took note of the fact that they had made many approximations in their calculations. Would their results stand up if more exact calculations were made? Wheeler wondered. Large computers were now available, making such calculations much easier.

Wheeler decided to redo the calculations, without the approximations, using the large calculator called MANIAC. He assigned the problem to two of his students, Kent Harrison and Masami Wakano. They began with the problem Oppenheimer had assigned to Volkoff. Doing away with the approximations he had used, they found that the result remained essentially unchanged. Then they went on to the Oppenheimer-Snyder calculation. Would the star still go into a never-ending collapse? When the calculations were complete, they found that it would; little was changed with the more accurate calculation. Wheeler and his students presented their results at the 1963 Texas Conference on Relativity in Dallas. It was the largest paper presented at the conference and was later made into a book.[8]

Wheeler's enthusiasm for black holes continued to grow, and he was eager to learn as much as possible about them. Incidentally, as mentioned earlier, he was the one who coined the name "black holes." Meanwhile, other groups were becoming involved in black hole research. Y. B. Zeldovich, I. D. Novikov, and others were making important breakthroughs in Russia, and a large group was forming in England.

COLLAPSE TO A BLACK HOLE

Let's look a little closer at how a black hole forms. Oppenheimer and Snyder's calculation showed that the collapsing star must have a mass greater than 3.2 solar masses if it is to form a black hole. But there are losses before the collapse and during it, so the initial star will likely have to have a mass of at least 8 solar masses to end as a black hole. Before we start, let's recap what we know about smaller stars. As we saw earlier, a star less than 1.4 solar masses collapses over millions of years to a white dwarf. Astronomers have observed hundreds of these white dwarfs in space and have learned a great deal about them. If the star is larger than 1.4 solar masses, but less than 3.2 solar masses, its final state is a neutron star, composed mostly of neutrons. Such a star is created when a star explodes as a supernova. It is the remnants of the explosion and is only a few miles across. Astronomers have observed many of these neutron stars.

We'll begin by assuming our star is nonspinning; in practice, most stars spin, but we'll deal with them later. When a star collapses to form a black hole, no explosion takes place. The star just implodes over a tiny fraction of a second. But why does such a star collapse? The reason is that in all stars there are two forces that keep the star in equilibrium. Pulling inward is the force of gravity; it is held in check by a pressure supplied by the nuclear furnace at the center of the star. This is the energy source that supplies the heat and light of the star throughout its lifetime. (Actually, some of the outward pressure is due to the gas pressure of the star.)

To generate its energy, a star needs fuel, and the fuel in this case is hydrogen. But eventually, as in the case of a car, the fuel is depleted. The star is then overcome by its gravitational field. If you could watch the collapse, you would see that it begins slowly, but quickly picks up speed, and then in less than a second it's all over. The star has become a black hole. Where there was once a massive star, there is now only a tiny black sphere in space; it is only a few miles across—a little smaller than a neutron star. But where a neutron star has a solid surface, the surface of the black hole is not solid. If you fell into it, you would pass through it without feeling anything. It would appear black because no light is emitted from a black hole. With no light being emitted, it might seem that you would not be able to see it, but this isn't true; it would block off background stars and would be seen as a black sphere.[9]

Although we wouldn't notice anything when we passed through it, a black hole does have a surface. It is referred to as the *event horizon*, and it is a one-way surface. You can pass through it going into the black hole, but once inside you can never get out again. The mass of the collapsing star is at the center of the event horizon; it is now a point mass and is referred to as a singularity. If you passed too close to this singularity, you would be pulled in and crushed.

The space around a black hole is severely curved. One way of checking this is to use a light beam. If you had a powerful searchlight and probed the region just outside the black hole, you would find that the light rays would be bent, depending on where you shone the light. At some distance from the black hole the beam would be only slightly bent, but as you got closer to the event horizon, the beam would be increasingly bent toward it, until at 1.5 times the radius of the event horizon, the light rays would be curved completely around the black hole. None would leave the region. Inside this, all beams would be pulled into the black hole.

THE NO-HAIR THEOREM

The black hole that we have been discussing is called a Schwarzschild black hole. Are there any other types? That depends on what remains after the collapse of the massive star. We know that the mass remains and that the gravitational radius depends on the mass: the greater the mass, the larger the gravitational radius and hence the larger the area of the event horizon. Furthermore, Werner Israel of Canada showed that all Schwarzschild black holes would be exactly the same. Even if the stars from which they were formed were quite different, there would be no way you could distinguish between the resulting black holes.

Are there any other properties of a star, besides mass, that remain after the star becomes a black hole? Scientists soon found that the spin of the star would remain. In short, a spinning star would produce a different type of black hole, as compared to a nonspinning one. Einstein attempted to find the solution to his equations for a spinning star, but failed. Finally, though, in 1963 Roy Kerr of New Zealand solved the problem, and the resulting black hole is now referred to as a Kerr black hole.

The Kerr black hole is quite different from the Schwarzschild black

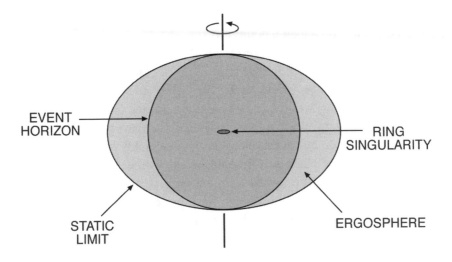

Fig. 25: Cross section of a Kerr black hole showing the ring singularity, ergosphere, event horizon, and static limit.

hole. It's well known that as a sphere spins faster and faster it becomes oblate (in other words, the distance across at the equator is greater than the distance across at the poles), and since the Kerr black hole is spinning, it is oblate. Like the Schwarzschild black hole, the Kerr black hole also has an event horizon, but beyond it is another surface known as the *static limit*. This is the position surrounding the black hole where a particle cannot remain at rest; it is pulled around by the spin of the black hole due to what is called *frame dragging*, a phenomenon that occurs close to any spinning object.

To understand the static limit a little better, assume you are in a rocketship approaching a black hole. Because of frame dragging, you will be pulled around the black hole in the direction of its spin. If you are at some distance from it, you can easily slow the rocket by using your retrorockets. But as you get closer and closer, it becomes increasingly difficult to slow your motion. Finally, at the static limit you are dragged around the black hole regardless of how hard you blast your retrorockets. You can't remain stationary once you are inside the static limit, but you can still escape from the black hole. You have to pass through the event horizon before you are trapped. The region between the event horizon and the static limit is called the *ergosphere*. And finally, at the center of the Kerr black hole is the singularity, but it is no longer a point; it is ring-shaped.

Another property that is preserved when a star collapses to a black hole is charge. And since charge produces an electric field, we can have a black hole with charge and an electric field. The solution for this type of black hole (and star) was found by Hans Reissner and Gunnar Nordström in 1917, so it is now called a Reisner-Nordström black hole. In practice, though, any charge that accumulates on a star quickly becomes neutralized, so this type of black hole probably doesn't occur in nature.

Are there any other properties that are preserved? As it turns out, only these three—mass, spin and charge—are preserved. This has become known as the *no-hair theorem,* which means that black holes cannot have hair—only the three properties above (another of Wheeler's puns). But this leaves us with one more type of black hole: one that is both spinning and charged. The solution for this type of black hole was found by C. T. Newman and several of his students at the University of Pittsburgh, so it is now referred to as a Kerr-Newman black hole.[10]

SPACE-TIME DIAGRAMS

One of the things we would like to be able to do is plot a trip in and around a black hole, and to do this we need what is called a *space-time diagram*. The idea behind such a diagram is simple; it is merely a plot of where we are in space at a given time. The two axes are therefore space and time. A simple space-time diagram is the plot of a trip between two towns. On this diagram you give your position at various times during the trip. The space-time diagram for a black hole is, of course, considerably more complicated because space is curved, and time does not run at the rate it does for someone on Earth.

Let's consider what we need to include in the diagram. We'll assume we are dealing with a Schwarzschild black hole, so it will have an event horizon and a singularity, and both will have to be included in the diagram. We'll select space and time as the two perpendicular axes, with space on the horizontal. We represent the singularity by a jiggly line and the event horizon by a dotted line (figure 26).

Now assume we are at some distance outside the black hole in a

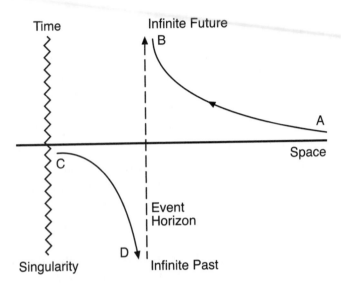

Fig. 26: Space-time diagram showing a trip into a black hole.
A trip from the singularity to the event horizon is also shown.

rocketship, and we release a probe that falls into the black hole. We'll assume there is a clock inside the probe that we can read. As we watch the probe from the safety of the rocketship, we see its clock runs slower and slower, and the probe never quite reaches the event horizon. We represent this by trip A to B in the figure.

We never see the probe enter the black hole, but if an observer were aboard the probe, he would notice that he passed through the event horizon in a relatively short time. Let's suppose therefore that we have an observer inside the black hole, near the singularity, at point C. Assume that he moves toward the event horizon in hopes of getting out of the black hole. As he tries, however, the event horizon will recede away from him, and he will never reach it. His trip is shown as C to D in the figure.

Scientists soon showed that this diagram is not the best one to depict the region around black holes. The reason is that it conceals certain features of the region. Just as a transformation to different coordinates showed us that the Schwarzschild singularity is not a real singularity, so, too, can a transformation help us here. The transformation that was needed was found by Martin Kruskal of Princeton University. Kruskal was a plasma physicist with little knowledge of general relativity and black holes. But he was curious, and his curiosity eventually got the best of him, so he decided to organize a study group to learn as much about them as possible. During his study he noticed a transformation that "unfolded" some of the regions of space and time in the above diagram. He was sure it was an important breakthrough, but when he showed it to Wheeler, Wheeler didn't seem interested. He was disappointed and decided not to publish it. Within a short time, however, Wheeler realized the discovery was important and began using it. He gave full credit to Kruskal.

At first glance, the Kruskal diagram looks quite different from the one above. There are now two singularities and two event horizons, but the space axis is still on the horizontal and the time axis on the vertical. Again, as in the previous diagram, we can show various trips into and around the black hole. We will discuss three trips, labeled as A, B, and C. They are shown in figure 27. Trip A is merely a trip between two points outside the black hole and is therefore of little interest. Trip B takes us

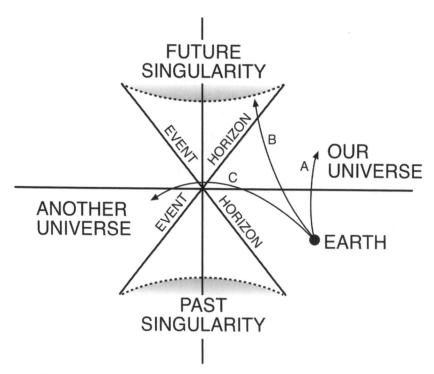

Fig. 27: Three trips in and around a black hole shown on a Kruskal diagram.

through the event horizon and into the singularity, so it is of some interest. Trip C appears to be the most interesting of the three because it takes us through the black hole to another region of space. What is this other region? We'll assume, as Einstein did, that it is another universe, although it could also be a distant point in our universe. When you look at this trip in detail, however, you see a problem. All space-time diagrams of this type are set up so that trajectories of forty-five degrees represent the speed of light, and this means that this trip requires a speed greater than the speed of light, so it is impossible. In short, we cannot reach another universe or a distant point in our universe using Schwarzschild black holes. This was the discovery that Einstein made back in 1935.

But what about the other types of black holes? We have three other types. As it turns out, their space-time diagrams are similar but more complicated. Since the Kerr black hole is the one most likely to occur in nature, we are most interested in it, and it is easy to show that you can pass through a Kerr black hole. In short, it is accessible, but as we will see later, many problems remain.

An important contribution to space-time diagrams was made by Roger Penrose of London shortly after the Kruskal diagram was discovered. In the Kruskal diagram there are four regions that extend off to infinity. Penrose showed that it was convenient to bring these four infinities in so that they were part of the diagram. His diagram is shown in figure 28.

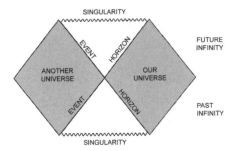

Fig. 28: A Penrose diagram.

THE PENROSE PROCESS AND MORE

In addition to his contribution to space-time diagrams, Roger Penrose made several other important contributions to black hole physics. His father was a professor of genetics, and his mother was a physician, and they hoped he would follow in the family tradition and become a medical doctor. Penrose thought about it, but it was soon obvious that he was gifted in mathematics, and over the objections of his family, he majored in mathematics at University College in London. It's perhaps a little ironic now that they worried that he wouldn't be able to get a job when he graduated.

He loved pure math and had no intention of getting involved in physics, but during his fourth year he met physicist Dennis Sciama. Sciama was impressed with his ability but wasn't able to persuade him to switch to physics. Penrose completed his PhD in mathematics but was eventually drawn to physics and began working on several problems in the area. He was on the train one day when he began thinking about what would happen to a particle that entered the ergosphere (the region between the event horizon and static limit) of a black hole. It would not be trapped in the region; it would have to pass through the event horizon before it would be prevented from getting out. Penrose looked into the case where the particle broke in two. He assumed one of the pieces escaped the black hole and the other passed in through the event horizon. To his surprise, he discovered that the exiting particle gained considerable energy; in fact, the energy it carried off was much greater than the energy the particle came in with. Where did the additional energy come from? Penrose soon discovered that it came from the black hole itself, which meant that the black hole lost energy as the particle exited. If the black hole was a Kerr black hole, this energy would come from the spin, and the black hole would spin slower. This meant that energy could be extracted from a black hole. The only thing that would be needed was particles that broke up inside their ergosphere.[11]

Black holes were obviously much more complex than previously thought. Indeed, in discussing them so far, simplifications have been applied. In practice, all black holes, except the Schwarzschild black hole, have two event horizons and two ergospheres—an inner and an outer one.

For a moderately spinning Kerr black hole, for example, the two event horizons are well separated. If we increase the spin, however, the two event horizons begin moving toward one another, until finally they merge. If we continue increasing the spin, the new double event horizon begins to shrink, until finally it disappears into the singularity. A similar situation occurs for a charged black hole.

PRIMORDIAL BLACK HOLES

The black holes we have discussed so far are referred to as *stellar collapse black holes*. They are created when a giant star collapses. Stephen Hawking of Cambridge University showed, however, that there is another type of black hole. Born in 1932 at Oxford, England, Hawking was the eldest of four children.[12] His father was a research biologist who hoped Stephen would follow in his footsteps. But Hawking had other ideas. It was obvious early on that he was good in mathematics and physics, but strangely enough, most of his teachers did not think he was an outstanding student.

He entered Oxford in 1959, and his genius soon began to flower. Despite having many outside interests, such as rowing, classical music, and science fiction, he excelled without working hard. He graduated with honors and went on to graduate school at Cambridge University. By now he had decided to become a cosmologist. Soon after entering graduate school, however, he began to notice that he was unsteady on his feet and was having trouble controlling his hands. He was soon diagnosed with the motor neural disease amyotrophic lateral sclerosis (abbreviated as ALS, and sometimes called Lou Gehrig disease after the American baseball player who died of it). When Hawking learned of his illness, he was devastated. He was sure that he was going to die. Indeed, he was so dejected he couldn't work and eventually gave up trying to finish his doctoral thesis. But his condition stabilized, and it was soon obvious that he wasn't going to die. This gave him a new lease on life, and he finished his thesis and got his PhD. He later married and had two children. Within a short time he was in a wheelchair, but the disease did not affect his mind, and over the next few years, he made a number of important discoveries.

One of Hawking's first discoveries was what we now call *primordial*

Fig. 29: Stephen Hawking.

black holes. While thinking about the early universe one day, he realized that the explosion that created the universe would not have been perfectly homogeneous. There would have been regions of over- and underdensity, and because of this, pockets of matter would have been compressed into black holes as compression and expansion occurred around them. In theory, a whole array of black holes would have been created, from tiny atomic-sized ones up to gigantic ones that may now reside in the cores of galaxies. The most interesting of these primordial black holes are the tiny ones, some of which are no larger than an atom. They are sometimes referred to as mini black holes. Space may, indeed, be filled with these mini black holes.

EXPLODING BLACK HOLES

After Penrose discovered a process for extracting energy from a black hole, a twenty-one-year-old graduate student, Demetrios Christodoulou, who was working for Wheeler at Princeton, became interested in the process. He wondered what would happen if a large numbers of particles were projected into the ergosphere of a spinning black hole, and each of them broke up, with one of the two exiting the black hole and the other falling into it. He quickly concluded that the black hole would eventually stop spinning. He referred to the remnant as the "irreducible mass."

When Hawking heard of his work, he became alarmed. He had just proved an important theorem in black hole physics, namely, that the surface area of a black hole never decreases. Furthermore, if two black holes merged, the overall area would be greater than the combined areas of the two holes. He realized that because of this, the surface of a black hole was similar to a concept in thermodynamics called *entropy*. Entropy is a measure of the disorder of an isolated system, and according to the second law of thermodynamics, the entropy of an isolated system always increases.[13] Furthermore, if two objects with specified entropies are brought together, the overall entropy of the new system will be greater than the sum of the entropies of the old systems.

Hawking, along with colleagues Brandon Carter and John Bardeen, showed that there was, indeed, an analogy between black hole physics and thermodynamics, but to Hawking that's all it was. It was only a coincidence. But he was soon in for a shock. About this time another of Wheeler's students, Jacob Beckenstein, began looking into the similarity, too, and he published a paper in which he stated that the surface area of a black hole was a direct measure of the black hole's entropy. It was not *like* entropy; it represented it in all respects. But there was a problem. Entropy, by definition, is associated with temperature, and the event horizon of all black holes had a surface temperature of 0K by definition. They emitted no radiation. Bekenstein therefore needed something to represent temperature, so he selected gravity.

Hawking was sure that the idea was wrong, but in 1973 he went to Moscow to visit Yakov Zeldovich and Alexi Starobinsky, two of the world's leading experts on black holes. To his surprise, he found that they were convinced that rotating black holes emitted radiation and therefore might have a temperature greater than 0K. He returned to England in a dejected state. It didn't make sense to him, but he began looking into the mathematics of the problem; indeed, he decided to apply quantum mechanics. This was particularly strange in that black holes were not generally thought of as quantum mechanical objects; quantum mechanics applies to atoms and molecules.

Hawking looked at the strong gravitational field near the event horizon. He knew that according to quantum mechanics space was not empty. What are called "virtual pairs" of particles (e.g., an electron and positron) pop into existence for a very short period of time, then disappear. They can do this because of what is called the "uncertainty principle," which tells us that there is a fuzziness associated with nature at the atomic level. This fuzziness conceals the pairs for a short period of time, and by the end of this time, they have recombined and disappeared. Hawking realized that pairs such as this would be produced in abundance just outside the event horizon because of the strong gravitational field there. He then considered the possibility that one of the two particles fell into the black hole and the other escaped. To his amazement, his calculations showed him that it was possible, and if this was the case, the black hole would appear to a distant observer to be emitting particles. Furthermore, radiation would also be emitted by the accelerating particles. But these particles and radiation carry off energy. Where did this energy come from? Hawking soon showed that it came from the black hole. In short, the surface of the black hole was radiating, therefore it was hot, and it was giving off energy. As a result of this energy loss, its surface area would shrink, and this in turn would increase its surface temperature, until finally it exploded in a tremendous burst of energy. The black hole would then be gone.

To Hawking's relief, he found that the surface temperature of ordinary stellar collapse black holes was only a tiny fraction of a degree above 0K, and their lifetime would therefore be extremely long—many times the age of the universe. But large amounts of energy would be released by tiny atomic-sized black holes, and their surface temperatures would be high. Furthermore, they would not last for a long period of time. Hawking showed that a black hole that formed in the big bang with a mass of 10^{15} grams would now be exploding. It would, of course, be much larger than an atom—about the size of a small mountain, when first formed.

When Hawking presented his results, many scientists thought he had gone off the deep end. Most were sure he had made a mistake. But when they looked at his calculations in detail, they found that he was right. Furthermore, he had made a fundamental breakthrough in the relationship between quantum mechanics and general relativity. He showed that the radiation emitted by the black hole satisfied the quantum formula for the emission of radiation derived by Max Planck in 1900. This was the first time a link had ever been found between the two major theories of physics.[14]

WHITE HOLES

As we saw earlier, when Einstein and Rosen were examining the space around a singularity in 1935, they discovered that it was like a bottleneck. It is now referred to as the throat of the black hole. They found that this throat narrowed down to the singularity, then opened again, almost as if a mirror-image throat was attached to the other end. We now refer to the overall object as a wormhole in space. Einstein puzzled over what was at the other end of the wormhole and came to the conclusion that there was "another universe" there. He was relieved when he was able to show that a speed greater than that of light was required to get through. But when the solution for a spinning black hole was discovered by Kerr in 1963, scientists realized that a person could pass through the wormhole with a speed less than that of light.

But if we can get through, we have to ask what we would see if we looked back at the exit. Clearly it can't be a black hole since black holes only pull matter in, and matter is pushed out at this end. Physicists now call this end a *white hole*. It is, in effect, a time-reversed black hole. To someone outside the region, it would appear to be a "gusher." Astronomers were excited when they heard about this, because there are objects in space that look like gushers, and this could be an explanation of them. As we will see, however, their hopes were soon dashed.

Do white holes exist? Einstein's theory of relativity shows that they can't be associated with stellar collapse black holes. If they do exist, they have to be associated with primordial black holes; in other words, they have to be created during the early stages of the big bang. But there is a problem. White holes would emit particles and radiation, and so would tiny black holes. Hawking has shown, in fact, that black and white holes would be indistinguishable at this time. And there is another problem: Doug Eardley of Yale University has shown that even if white holes began to form in the early universe, black holes would quickly form around them, leaving only black holes.

So we have a problem. It seems that black holes have a wormhole associated with them, and matter can be dragged in through this wormhole. And it is possible that some of this matter can avoid the singularity and escape through the other end of the wormhole. But the other end, namely, white holes, may not exist.

INTO A BLACK HOLE

Do black hole wormholes actually take us to distant points of the universe? If so, they are exactly what we have been waiting for. We know that it is almost impossible to travel between the stars by conventional means because we are restricted to velocities less than that of light, and most stars are many light-years away. The times involved would be too long. Black hole wormholes may be a way around this, so it's natural to ask: what would it be like to enter a black hole wormhole? Several people have made computer calculations of such trips.[15] In practice, for us, it would probably be best at first to send a probe into the black hole. It could send back information and tell us about the problems that we would encounter. We already know, however, that there would be many problems. Let's look at some of them.

First of all, it has been shown that black holes are unstable. The throat of a black hole pulses; in other words, it opens and closes. In particular, anything that enters the throat causes it to pinch off. If a rocketship, for example, entered one, it would soon be caught in the pinchoff and would be crushed. Furthermore, a black hole event horizon is one-way, so if you passed through it, you could never get back through it. This means that wormholes are also one-way. You could pass through one and perhaps end up at some distant point in the universe, but you could not come back

Fig. 30: A simple representation of a wormhole in space.

through the same wormhole. You would need a different one, and it's quite unlikely you would find another one near the exit of the first one.

Earlier we discussed another problem referred to as tidal forces. If you were falling toward a black hole feetfirst, the gravitational force on your feet would be much greater than on your head. As a result, your body would be stretched, and the closer you got to the black hole, the greater the stretching. For most black holes, you would look like a piece of string by the time you got near the singularity. As it turns out, there is a way around this. Very massive black holes exert only small tidal forces, but this requires a black hole with a mass thousands of time greater than our sun. The only place we see such masses (and possibly black holes) is in the cores of galaxies.

Next, we have a radiation problem. Hawking showed that particles are ripped out of the vacuum near the event horizon of a black hole. For average-sized black holes, this is not a serious problem, but the space near the singularity of a black hole is extremely warped, and a considerable amount of radiation and particles would be produced. It would, in fact, be so great that it's unlikely we could pass through this region.

Finally, we have the difficulty with exits. We saw that white holes may not exist. How would we get out of the wormhole? And there are other problems: how, for example, would we get by the singularity? We would obviously have to avoid it; otherwise, we will be pulled in and crushed. In the next chapter we will see that there are also serious "causality" problems. Causality is the idea that the effect of something has to come after the cause.

EINSTEIN'S INFLUENCE

It's no secret that Einstein didn't like the idea that black holes might exist. He struggled hard to show that they couldn't exist. He thought of them as a flaw in his theory—a shortcoming. Once he realized they were a natural outcome of the theory, he may have been less adamant about their existence, but we can't say this for sure, since he said nothing about them during the last few decades of his life. Despite his reluctance to accept them, however, there's no doubt that his theory has had a large influence on them. Black hole physics is now a major branch of theoretical physics, with hundreds of physicists around the world working on various problems related to them. And all of them are using Einstein's general theory of relativity as their major tool. Although black holes are predicted in other theories, even Newton's theory, they do not explain them the way general relativity does.

Indeed, not only has black hole research developed theoretically, but there are now numerous astronomers looking for black holes in space. And over the past few decades, many excellent candidates have been found. One of the most famous is called Cyg X-1.[16] Many astronomers believe that huge black holes may also reside in the cores of galaxies, and there are several ongoing searches for them.

Chapter 4

The Mystery of Time and Time Travel

Two of the staples of science fiction are time travel and travel to the stars, and it goes without saying that in real life neither of these will be easy. Our main problem in relation to travel to the stars is that we are restricted to the speed of light, and most stars are many light-years away. We would obviously have to travel at speeds very close to that of light if we were to reach them in our lifetime. And this isn't the only obstacle; a simple calculation shows us that we would need an incredible amount of fuel. Its weight would be millions of times that of the rocketship. A way around this would be to take a shortcut through space, and as we saw in the last chapter, black holes might allow us to do this. But we also saw that there are many challenges, and it appears that we may not be able to overcome them.

Fortunately, there are alternatives, but before we look at them, let's take a close look at time. After all, if we want to travel to the stars in a reasonable amount of time, we're going to have to understand time. We begin by asking: what exactly is time? If you asked several people that question, you would no doubt get a different answer from each of them.

And the truth is: nobody really knows. It's easy enough to point to your watch and say, "It's what that thing is measuring." But does that really tell us anything? No, it doesn't, and even if you quizzed scientists, they would admit that they don't understand time any better than anyone else. They deal with it in their equations, but it's just a variable they manipulate when they work with these equations. Time is, indeed, a mystery. And, as we'll see, it's more of an enigma than you'll ever find in any mystery novel.

Our concept of time has changed over the years. Newton was one of the first to try to define it. He said it was universal and absolute; in other words, it was the same everywhere throughout the universe, and there was no way we could ever change it. But when you think about this for a moment, it's easy to see that it's flawed. The speed of light causes problems. We are restricted to it, and because of this, we can't say what is happening at any given time at two different points in our universe. If we had an observer in a rocketship near Jupiter, for example, we can't say what is happening in it right now. It takes light roughly a half an hour to reach Jupiter from Earth, so the best we can do is to say what happened a half hour ago. The rocketship could have exploded twenty minutes ago, and we wouldn't know it for another ten minutes.

No one challenged Newton's idea of an absolute time throughout the universe for over two hundred years. Then came Einstein. Not only did he show that Newton was wrong; he showed he was wrong for two different reasons. First, he demonstrated that the rate of passage of time would be different for two observers in motion. In short, they would not see time pass at the same rate. Then a few years later he showed that the rate of passage of time also depended on the strength of the gravitational field the observer was in.

But even Einstein, who knew a lot more about time than most people, was confused by its true meaning. He once said to a friend, "The past, present, and future are only illusions, even if stubborn ones."[1] That may sound a little strange coming from someone who revolutionized our ideas about time. Most people don't believe that the past, present, and future are an illusion. They can remember the past; they can pinch themselves in the present, and they have hope for the future. To them, it seems as if time *flows* like a river.

THE RIVER OF TIME

Is time really like a river? Does it flow? It doesn't take too much thought to see the shortcomings of this idea. After all, if it were a river, it would flow at one rate for all observers, and because of Einstein, we know that this isn't the case. Two observers that are moving with different velocities see it pass at different rates. Furthermore, as we just saw, a gravitational field also affects the passage of time.[2]

The major problem with the idea of the flow of time is that we can't compare it to anything. We can't simply say that it's flowing at one second per second; that doesn't make sense. Time has to be judged against something external, something more fundamental, and there is nothing. There's only time. It simply exists; it doesn't flow. In fact, there's nothing in the laws of physics or the basic equations dealing with time that says anything about the flow of time. How, for example, would we measure a sudden increase in the rate of flow of time? All our clocks would undergo the same change, so we would be measuring it with a clock that was also running faster.

Another question we might ask is: can we reverse time? In other words, can time run backward? For most of us, that would be nice—at least for a while. We would get younger. Unfortunately, there's no way we know of for reversing time. If we reverse time in the equations of physics, the result is the same. So physics is not affected by a time reversal. Time does not flow in either direction, but there is a direction to time. Intuitively we know that events are ordered. Certain things happened yesterday and the day before, and we remember them, and we feel we have a sense of "now." So there is, indeed, an *arrow of time.* The arrow isn't moving, but it points. Actually, as we will see, there are several arrows of time.

An arrow of time is something that gives us a sense of past, present, and future. The best-known arrow of time comes to us from thermodynamics, the study of heat in motion. To understand it, we have to go back to a concept introduced earlier, called entropy. I mentioned that it was a measure of the disorder of a system. According to the second law of thermodynamics, the entropy of an isolated system must always increase. We can illustrate this with a jar of marbles; let's assume that they are arranged so that the bottom half of the jar contains all red marbles, and the top, all blue marbles. At this point our system has minimum entropy; in other words, the amount of disorder is least. Now, begin shaking the jar. As time passes, more and more blue marbles will move from the top half into the bottom, and red marbles will move to the top. As we continue to shake the jar, the red and blue marbles become more and more mixed, or disordered. This means that the entropy of the system has increased, and it's pretty easy to see that it's never going to decrease. For this to happen, the red and blue marbles would have to separate again, and the probability of this happening is extremely low. It's important, of course, that our system be isolated. If we poured a few blue marbles into the top half, the entropy would decrease, but our system wouldn't be isolated.

Since we know that the entropy of an isolated system always increases (and therefore distinguishes the past from the future), we can use it as an arrow of time. The increase in entropy points in a particular direction, namely, the future.

Another arrow of time comes from the expansion of the universe. As we look out at the galaxies today, we see that they are all moving away from us. This means that in the past they were closer together, and in the future they will be farther apart. The expansion of the universe can therefore be used to give direction to time. A third arrow comes from our physiological sense of the passage of time; we remember the past and look forward to the future. This arrow may not be as strong as the other two because it's based on human intuition, but it is an arrow nevertheless.

Other arrows of time exist, but they are a little more subtle.[3] One comes from quantum mechanics, the theory of the atom. A quantum system loses information when it is probed—in other words, when we make a measurement—and this loss points in a particular direction. In the same vein, there is also an arrow of time associated with matter-antimatter annihilation. For some unknown reason, it is slightly more likely for antimatter to be converted to matter than vice versa, and this gives direction to the reaction.

EINSTEIN'S TIME

Let's take a closer look at what Einstein showed us about time in his special theory of relativity. Published in 1905, it revolutionized the way we look at both time and space. According to the theory, the rate of passage of time depends on the motion of the observer. To illustrate, assume we have an observer on Earth and another one in a rocketship passing Earth. We will assume that they synchronized their clocks before the rocketship took off. As the rocketship passes overhead at greater and greater speeds, the time on its clock, as seen by an observer on Earth, will appear to go slower and slower. And indeed, if he lands back on Earth, a shorter interval of time will have passed on his clock than on the clock on Earth. Strangely, the observer in the rocketship will not notice anything odd; time will appear to pass normally for him.

If the rocketship could travel at the speed of light, its clock would appear to stop, according to the observer on Earth. As Einstein showed, however, the speed of light is the uppermost velocity allowed in the universe, and matter cannot attain it, so the clock never stops. Incidentally, if the observer in the rocketship looks back at the clock on Earth, he will see it run slow. So the effect is reciprocal.

Einstein also showed that time is affected by a gravitational field. According to his general theory of relativity, a clock in the basement of a skyscraper runs slightly slower than one in the penthouse. The difference in this case is extremely small because the difference in gravitational field intensity is small. But there are places in the universe where the gravitational field changes dramatically over a short distance. One of these is near a neutron star. Neutron stars are particularly dense, and they are only a few miles across. Because so much matter is packed in such a small space, their surface gravity is extremely high, but as you move outward from the neutron star, the field drops off rapidly. This means that if we took our two observers from Earth (along with their clocks) to a neutron star, we could see a big difference in the rate at which their clocks run. Suppose one of the observers ventured to the surface of the neutron star, and the other stayed some distance away. We'll assume they use telescopes to look at each other's clocks. The observer out in space will see the clock of the observer near the neutron star run much slower than his. Over a period of twenty-four hours on his clock, he would see only about sixteen hours pass on his friend's clock.

Things would be much more dramatic if we considered a black hole. Again, we'll assume we have the same two observers. Both are initially out in space at some distance from the black hole. One of the observers then moves closer to the black hole, and as he does, the distant observer sees his clock run slower and slower. Indeed, if the first observer went all the way to the black hole's event horizon, the distant observer would see his clock stop. Gravity is so strong at the surface of a black hole that it stops time. Unlike the case with special relativity, however, there is no reciprocity in this case. When the observer near the black hole looks back at the outside observer's clock, he sees it speeded up.

IS TIME TRAVEL POSSIBLE?

Travel to the stars is, of course, common in science fiction. Travel to the past and future are also used extensively. But are such things really likely to occur in the future? For years scientists were skeptical. Mysterious "stargates" were usually used to travel to the stars, and to the future and the past, in science fiction, but the science behind them was never explained. After black holes were predicted, they caused a sensation in the world of science fiction and were frequently used as stargates. But scientists knew they wouldn't work. There were too many conflicts with traditional science. If someone in a rocketship decided to dive into a black hole in the real world, she would almost certainly die. If the tidal forces were low enough to get close to the black hole without being pulled apart, she would be killed by the radiation or crushed by the collapsing throat of the black hole. This didn't seem to worry science fiction writers, however; they needed to move their characters around the universe, and they had to have something.

For the most part, scientists ignored the misconceptions. Science fiction was popular, and without time travel and so on, it would hardly be science fiction. Indeed, some of the writers were scientists themselves. One of these writers was Carl Sagan of Cornell University.

SAGAN AND THORNE

In 1984 Carl Sagan wrote a novel that he titled *Contact*.[4] In the novel he had the heroine, Ellie Arroway, enter a black hole near Earth, pass through its wormhole, and emerge near the star Vega, about twenty-six

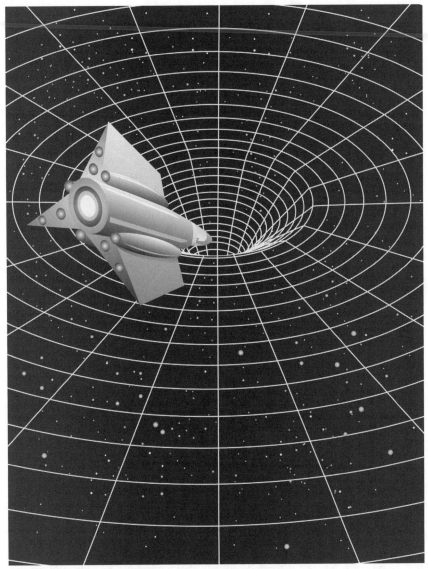

Fig. 31: A spaceship entering a wormhole in space.

light-years away. Sagan was worried after he wrote the novel, however. As a scientist, he was sure it would be a bad reflection on his reputation if he used bogus science in the novel. He wanted the science to be correct so that events in it would be possible in the distant future. He knew little about black holes and space-time tunnels, but his friend Kip Thorne of Caltech was an expert in the area. He therefore sent his manuscript to Thorne to look over its scientific accuracy.

Thorne read the book and was disturbed. He knew there was no way a black hole could be used for space travel, but he wanted to help Sagan. He wondered about the alternatives. If a black hole wouldn't work, was it possible that a wormhole would? Isolated wormholes didn't exist in the universe, as far as he knew, but they were worth looking at. Thorne wrote down Einstein's equations and began looking for wormhole solutions, and to his surprise, he found one. Furthermore, under the right conditions, it could be used to traverse long distances in space. He saw, however, that converting the wormhole to a usable subway through space would not be an easy task. Thorne outlined the problem and turned the details over to his student, Michael Morris.[5]

Before we get into the details, let's look at how a wormhole might be formed. We'll begin by representing our three-dimensional world by a two-dimensional sheet. As it turns out, this is the only way of illustrating wormholes. The sheet is in hyperspace, a region we cannot see or detect. Now, assume that this sheet becomes warped by a singularity, and at the same time another nearby sheet of space becomes warped by another singularity. Finally, the two singularities somehow find one another in space and annihilate each other, and in the process they form a wormhole between the two sheets. This is the type of wormhole Thorne envisioned, but as you can probably guess, the probability of this actually happening is extremely low. Nevertheless, this wormhole had many of the properties needed for travel through space.

Let's look at what is needed. First of all, pinchoff had to be eliminated. Second, tidal forces had to be small, and third, the wormhole had to be two-way, which meant there could be no event horizon in it. Furthermore, transit times had to be reasonable, and finally, radiation effects had to be minimal. As it turned out, all of these would be satisfied by the wormhole, except pinchoff. Thorne and Morris considered the possibilities. One of the most reasonable was that the wormhole could be lined with some sort of material that would withstand the pinchoff. They made some calculations. The inward pressure would be so large there was nothing on Earth that could withstand it. Finally, they realized that there was a type of material that might work, namely, negative mass, or negative energy material. It would have antigravity and would exert an outward force. Calculations soon showed that it would work, if it could be produced in sufficient amounts.

But where would they find material with negative mass? Thorne remembered that Hawking had predicted such material in the mid-1970s. Indeed, he had showed that the creation of virtual pairs just outside the event horizon of a black hole would cause the event horizon to shrink. According to Hawking, some of the particles and antiparticles would fall into the horizon and some would escape, and the ones that went into the horizon would cause a flow of negative energy into the black hole.

This was encouraging; furthermore, it wasn't the only evidence that negative energy material could exist. In 1948 the Dutch physicist Hendrik Casimir showed that negative energies could be created between two conducting metal plates.[6] If two such plates were placed in a vacuum, so that the energy between them was initially zero, short-lived particles and antiparticles from the vacuum would produce negative energy as the plates were attracted toward one another. All in all, then, the evidence for negative mass was strong.

It's important to point out that negative mass is not the same as antimatter. When matter and antimatter come in contact with one another, they annihilate each other with the release of gamma rays, and gamma rays are a form of positive energy. If negative matter were antimatter, the annihilation would produce nothing. Indeed, antimatter is attracted gravitationally to matter, but negative mass is not; it is repelled.

Thorne began calling the new material "exotic matter." He finally came to the conclusion that if you could thread the wormhole with exotic matter, it could be held open. But it would take a lot of exotic matter, which concerned Thorne.

In summary, then, a wormhole would have two mouths, each in ordinary three-dimensional space, connected through hyperspace by a throat that is propped open by exotic matter. The two mouths would be only a short distance apart through hyperspace, but they might be light-years away from one another in ordinary space.

This brings us back to the problem: where do we get the wormhole? Thorne showed that there are three possibilities. They could be formed in the big bang explosion, but this seems unlikely, and any that may have formed would likely have disappeared long ago. The best possibility was

Fig. 32: A scientist waving at an astronaut who is
thousands of light-years away, through a wormhole in space.

associated with quantum theory. In the mid-1950s John Wheeler showed
that, just as virtual particles pop in and out of existence in the vacuum, so,
too, do wormholes pop in and out of existence on a much smaller scale.
On a scale trillions of times smaller than atoms, Wheeler believes there is
a "quantum foam." It is like a froth of bubbles, but it contains wormholes.
According to Thorne, a supercivilization may be able to dip down into
this foam, pull out a wormhole, and enlarge it. It would no doubt be a difficult process, but it's not impossible.

The only alternative, it seems, is to create one on a macroscopic level
by warping, or curving, space, and this would likely be even more diffi-

cult than getting one out of the quantum foam. We have no idea how we would do it.

TRANSFORMING TO A TIME MACHINE

Now, let's assume that we have a wormhole that can take us to distant points in space. Would it be possible to make it into a time machine? Could we twist the hole so that the second mouth ended up close to the first, enter the first mouth, and emerge out of the second one in the past or the future? Again, it sounds like something from science fiction. We will see that it is, indeed, possible, but there is a problem. According to the laws of physics, we can only go back into the past to the time when the machine was made. It's impossible to go back any further, which is, of course, a disappointment. On the other hand, we might be able to find a time machine that was made by a civilization that existed many years ago. Although this is highly unlikely, it is an interesting possibility.

Scientists have shown that there are two ways a wormhole can be converted into a time machine. The first involves the time dilation of special relativity. We know that if an observer moves at a speed close to that of light relative to us, time on his clock will pass slower than ours. Because of this, if we take one of the mouths of the wormhole and transport it via rocketship through space at speeds close to that of light, the wormhole will be a time machine when we bring it back to Earth. If we pass through it in one direction, we will go to the past, and in the other direction, we will go to the future.

A better, or at least easier, method of obtaining the same result was pointed out by Igor Novikov of the Space Research Institute in Moscow and Valery Frolov of the Physical Institute of Moscow in the late 1980s.[7] They showed that if you placed one of the mouths in a strong gravitational field, the wormhole will become a time machine. Indeed, the stronger the field, and the longer you leave it in the field, the greater the effect. In this case, we would take one of the mouths close to the surface of, say, a neutron star, and leave it for a while. A black hole would be more effective, but it might be difficult to deal with. Again, if you brought the mouth back to Earth and placed it close to the first mouth, the time dilation would remain, and the wormhole would be a time machine.

Fig. 33: A wormhole in space. Time is affected as you move through it.

OTHER METHODS FOR TIME MACHINES

Wormholes are perhaps the best way to build a time machine, but other methods have been discovered. What we need is referred to as a "closed time loop." In this loop you travel forward in time (according to your local clock), but in the process you move along a path in warped space-time that takes you to the past. Time is, in essence, bent around in a circle, bringing you back to the point where you left, but in the past.

Are there solutions to Einstein's equations that describe closed time loops? Indeed, there are. The first was discovered by W. J. van Stockum in 1937. He showed that closed time loops would exist in the region surrounding a dense, infinitely long spinning cylinder. It would, in effect, act as a time machine. But because it was infinitely long, it was not taken too seriously. Then in 1949 Kurt Gödel, a mathematician at the Princeton Institute for Advanced Study, discovered that a rotating universe would contain closed time loops. Einstein, who was also at the institute, was disturbed by the discovery but soon decided it was of little consequence, since our universe wasn't rotating.

In 1974 Frank Tipler of Tulane University picked up on van Stockum's discovery.[8] He showed that the cylinder need not be infinitely long. His cylinder was 100 km long by 10 km wide (62 miles by 6.2 miles). But he soon ran into problems. It turned out that it was unlikely the cylinder could be made strong enough to withstand the tremendous strain it would undergo when spinning fast enough to become a time machine. According to Tipler, all you had to do to use it as a time machine was orbit it a few times, and you would go back to the past. Science writer John Gribbin showed that Tipler's cylinder would need to have the mass of our sun, and it would have to rotate at half the speed of light if it was to work as a time machine.[9] It would obviously not be easy to achieve either of these.

CAUSALITY AND PARADOX

Time machines are, indeed, an intriguing possibility. What would it be like to go back to the past and visit yourself? It doesn't take much thought to see that this presents a number of paradoxes. The best known of them is referred to as the grandmother (or grandfather) paradox. It is as follows: Suppose you went back to the past and encountered your grandmother as a young lady, before she married your grandfather. For some unknown reason (I won't speculate on it), you decide to do away with her. But if she dies before she marries your grandfather, how could your mother (or father, whichever) have been born, and how could you be here?[10]

There are, indeed, several paradoxes of this type. Another one is as follows: Suppose you went into one of the mouths of the time machine and returned via the other before you left. You could go over and say hello to yourself. This sounds a little crazy, and indeed this poses a problem. Why didn't you see yourself greeting yourself the first time through? You obviously can't have things both ways; either you saw yourself the first time or you didn't. To make things a little more explicit, let's assume you hit yourself on your head with a stick, and it leaves a large bump. You would, of course, come back with a bump. Then, when you went over to the first mouth to hit yourself, you would already have a bump from the first hit. The bump obviously can't appear, then later you hit yourself. The problem here centers on what is called *causality*. According to this principle, as we saw earlier, cause has to come before effect, and this can obviously get mixed up in a time machine. Did you go into the time machine the first time with a bump on your head? If not, you can't have it the second time. Everything has to be logical, and causality has to be satisfied.

We may be able to observe the past in a time machine someday, but it's obvious that we won't be able to disturb it. If so, we would have two different outcomes to an event. Strangely enough, scientists have considered this. In 1957 Hugh Everett of Princeton University published what he called the "many world" interpretation of quantum mechanics. According to his interpretation, it is possible for the universe to break off into *parallel universes,* according to the choices that are possible at a given event.[11] This means that if you went back into the past and killed your grandmother, the event would take place in a parallel universe, not the universe we are familiar with. We will have more to say about parallel universes later.

PROBLEMS

Let's take a closer look at some of the difficulties of making a time machine. It's obviously not something that's going to be accomplished in the near future. Indeed, the problems are so severe that it will take hundreds, and likely thousands, of years to overcome them. Two of the most difficult problems are obtaining a wormhole and obtaining exotic matter to stabilize it. I mentioned earlier that scientists believe that the best place to get a wormhole is out of the quantum foam. It seems reasonable that this foam exists, but we don't know for sure that it does, and if we are to understand it, we need a quantized version of gravity, and so far this has eluded us. The two great theories of physics at the present time are quantum mechanics and general relativity. One governs the very small, and the other the very large (and in between). But the theories are radically different, and no one has ever found a connection between them (with the exception of the slight connection that Hawking found in relation to radiating black holes). What is needed is a theory of quantum gravity, which is, in effect, a unification of general relativity and quantum mechanics. It would explain the quantum foam and bring the two theories together. Considerable work has gone into this problem, with much of the recent work being done in connection with string theory. I will discuss string theory in later chapters.[12]

But even if the foam exists and we manage to isolate a wormhole, we still have to bring it up to reasonable size. This wormhole is trillions of times smaller than an atom and would require an enormous expansion. How would we do it? At the present time we have no idea. We would need something like the inflation that occurred in the early universe, but how would we create or control it, and how could we stop it when the wormhole was the right size?

Another problem is the exotic matter that would be needed to stabilize it. A number of physicists have shown that a tremendous amount would be needed, even for a moderate-sized wormhole. The recent discovery of the accelerating universe gives us a little bit of hope in this direction. According to this discovery, 70 percent of the universe is composed of negative energy. A number of scientists do not believe that this is the same negative energy that we would need for a wormhole, but its existence indicates that we still have a lot to learn, and, indeed, one day we may discover a good source of negative energy.

Finally, we have a problem related to the rays of light that pass through the wormhole. Thorne showed that because of the antigravity associated with exotic matter, the rays would diverge as they moved through the wormhole. Furthermore, several physicists have shown that if the two mouths were placed close together, light from one would enter the other, and if it did, it would be amplified. Indeed, this would take place again and again until the energy buildup was so large that the wormhole exploded. Thorne is convinced, however, that a way will be found around this problem. Actually, Tom Roman of Central Connecticut State University has found a way. He has shown that the problem can be overcome by using two wormholes, but there are difficulties with his method.

On the positive side, Lawrence Ford, also of Central Connecticut State University, and Tom Roman have found that negative mass may be helpful for more than stabilizing wormholes.[13] They showed that it can be used to create warp speeds, in other words, speeds greater than that of light. Warp speeds are, of course, another staple of science fiction.

WHERE ARE THE VISITORS FROM THE FUTURE?

Let's assume that sometime in the future we are able to make time machines that take us to the past and the future. If so, there would be civilizations out there that are well advanced beyond ours that have built time machines. And they would certainly use them to go to the past. Why aren't we seeing any of these visitors? You might argue that UFOs are from the future of Earth. But there are few indications that this is the case.

There are several reasons why they might not be visiting us. First of all, they might not exist. Second, as I mentioned earlier, a time machine can take you back only to the moment it was created, so if it is to take them back to our civilization today, it would have to had been created now. If an advanced civilization on Earth created a time machine, they would most likely have done it in the future, and therefore can't reach us. On the other hand, it may be that the laws of physics forbid travel to the past; Hawking has put forward what he calls his *chronology protection hypothesis* (the idea that the laws of physics conspire to prevent time travel by macroscopic objects), and maybe it is valid.[14] There's also the possibility that the idea of parallel universes is valid, and they are visiting universes parallel to ours, but not ours.

This, of course, brings us to the question of the likelihood of super-civilizations.[15] At the present time we have no evidence for them, but it is possible that they exist. It took us only about five billion years to develop life and technology, so we had a rather late start. It's reasonable to assume that most civilizations started much earlier than we did, and, if so, their technologies would be much more advanced than ours. With two hundred billion stars in our galaxy alone, this seems reasonable. Some of these civilizations may, of course, have lasted for a few thousand years or less.

EINSTEIN'S VISION

Time was an enigma when Einstein considered it back in 1905 and 1915, and it's still an enigma today. It's like light; we know how light behaves, and we can make all kinds of predictions about it, but no one really knows what it is. In the same way, no one can say they thoroughly understand time. But we do know a lot more about it than Newton did, and this is primarily due to Einstein. He was the first one courageous enough to break away from the stranglehold of Newton's ideas. No one dreamed that time could pass at a different rate for different observers, but Einstein showed it was possible, if the two observers were in relative motion, or if one of them was in a much stronger gravitational field than the other.

We are still struggling to understand time, but no major break-throughs have occurred since Einstein's time.

Chapter 5

Ripples in the
Curvature of Space

S hortly after Einstein published his general theory of relativity, he began wondering how he could use it to solve some of the outstanding problems in physics and astronomy. What would it predict? It was, after all, a new window to the universe, and there was no doubt that much could be learned by using it. One of his first applications came in 1916. It was well known that oscillating charges gave rise to electromagnetic waves. Was it possible that the oscillating masses would also give rise to waves? They would be weak because gravity was a weak force, but they should exist.

Using the equations of general relativity, Einstein looked into the problem and found that gravitational waves should, indeed, exist, and he obtained a formula that described them. He published his results in June 1916, but he wasn't satisfied with his calculation, and after checking it through, he found he had made a mistake. He therefore published a second paper in 1918 correcting the mistake and extending the ideas of his previous paper. It would become a classic paper on gravitational waves.[1]

Einstein found that gravitational waves would be given off by matter when it was accelerated. He considered a rotating rod, but he was disappointed to find that the intensity of the gravitational radiation from it was exceedingly low. He saw that it would take a much more massive system if the waves were to be detected. He therefore considered a binary star system (two stars revolving around one another), but was disappointed again. The mass of the stars would have to be much greater than that of ordinary stars if the radiation was to be detectable. His calculations showed that the stars would have to be exceedingly dense and close together, and at that time no such stars were known. Einstein eventually became convinced that gravitational waves would never be detected.

Several years later, in the mid-1930s, Einstein came back to the gravitational wave problem. He had used several approximations in his 1918 paper and wanted to derive a more accurate formula. By this time he was in the United States, so it seemed natural to him to submit his paper to *Physical Review*. To his surprise, however, he got it back stating that it was not acceptable in its present form. The paper had been sent out to an anonymous referee who had made a list of suggested changes that would make it acceptable. Einstein was outraged. He had never had a paper rejected before, even when he was unknown, and he was now a world-famous scientist. He refused to resubmit it. Eventually, however, along with Nathan Rosen, he began looking through the derivation again, and, to his surprise, he found an error. So it was perhaps fortunate that it wasn't published. In 1936 he and Rosen corrected the mistake and were able to derive an exact formula. It was published in the *Journal of the Franklin Institute* in 1937.[2]

Gravitational waves are, without a doubt, exceedingly weak, but there's no questioning that they exist. A simple argument shows why. Suppose an asteroid suddenly struck Earth; it would add to Earth's mass, so Earth's gravitational field would increase. The increase would be extremely slight; nevertheless, at a given point near Earth in space, there would be a slight increase in the gravitational field. We know that this can't take place instantaneously, and therefore a disturbance of the field must propagate out from Earth and make the change. This disturbance is a gravitational wave.

Let's compare gravitational waves to electromagnetic waves (figure 34). There are, indeed, many similarities, which will help us understand gravitational waves better. Electromagnetic waves (such as radio waves)

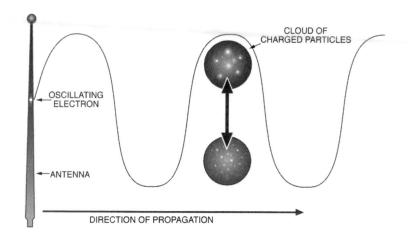

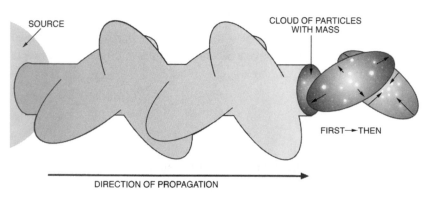

Fig. 34: Comparison of an electromagnetic wave (top) and a gravitational wave (bottom). The electromagnetic wave is a dipole wave; the gravitational wave is quadrapole.

are produced by accelerating charges; electrons rushing back and forth on an antenna produce the waves. This is how we receive the signal from a radio or television station. These waves have a particular wavelength and frequency; furthermore, they carry energy, travel at the speed of light, and vibrate in a particular way. The vibrational mode is referred to as the wave's polarization. For a simple antenna, the electromagnetic waves are emitted in a single plane and are called dipole waves.

Gravitational waves are similar in that they are given off by accelerating masses (rather than charges). They are much weaker because the gravitational field is much weaker than the electromagnetic field. The gravitational field of an electron, for example, is negligible compared to its electric field. This means that a tremendous amount of mass is needed to produce detectable gravitational waves. In particular, the mass must be concentrated, or dense, and because of this, neutron stars and black holes are the best sources.

Like electromagnetic waves, gravitational waves also have a wavelength and frequency, and they carry energy. In addition, they travel at the speed of light and have a polarization. Simple dipole gravitational waves, however, do not exist; the simplest gravitational waves are quadrapole, which means that there are two planes of vibration perpendicular to one another. These planes are also perpendicular to the direction of travel of the wave. To see how these waves affect a distribution of mass, consider a wave that is passing through a cloud of particles. As the waves passes through the cloud, the particles to the right and the left of it will move away from it, while those above and below will move toward it. Moments later the two roles will be interchanged.

WEBER

For many years after Einstein predicted gravitational waves, there was little interest in them. They were exceedingly weak, and no one thought they could be detected. In the late 1950s, however, Joseph Weber of the University of Maryland decided to try his hand at detecting them. Born in Patterson, New Jersey, in 1919, Weber obtained an engineering degree from the Naval Academy in 1940. After he graduated, the United States entered World War II, and he enlisted in the navy. He was assigned to the aircraft carrier *Lexington* as an electronics expert in radio, but within a short time, the *Lexington* was sunk and he was reassigned to a submarine chaser.[3]

Weber learned a lot about electronics during the war. His expertise in the area was so great, in fact, that after the war he was hired as a full professor of electrical engineering by the University of Maryland. He was asked, however, to work on a PhD while he was teaching, and he therefore signed up at the nearby Catholic University of Washington. In 1951 he obtained a PhD in physics. It's interesting that while working on his doctorate, he came up with the principle of the maser (forerunner to the laser), but he didn't have the resources to build a working model, so the honor of inventing the device went to others.

In the early 1950s Weber became interested in general relativity, and when he was eligible for a sabbatical in 1955, he decided to use the time to study general relativity. He went to Princeton University to study under John Wheeler, and while there, he talked to Wheeler about searching for gravitational waves. With his expertise in electronics and instrumentation, he was sure that he would have a good chance of success. Wheeler encouraged him, but there was a problem: no one had ever tried to detect the waves, so he would have to start from scratch. It soon became obvious to him, however, that there were several ways he could proceed. Three types of detectors appeared to show promise: a free-mass detector, an almost-free-mass detector, and a resonant detector. The Earth-Moon system is a good example of a free-mass detector. If a gravitational wave passed through them, both would oscillate slightly, but Weber soon realized that the oscillations would have to be measured to an accuracy of one part in 10^{17}, and there was no possibility of designing a detector with this degree of accuracy. The second type of detector would require two or more large suspended masses, with a separation detector placed between them. When a gravitational wave passed through this system, the masses would undergo slight oscillations that could be measured. But, again, there were difficulties.

Weber decided to use the third type of detector, the resonant detector, and he settled on an aluminum cylinder as an antenna. When a gravitational wave passed through it, it would vibrate slightly in tune with the frequency of the wave. But there was still an obstacle: how could you detect the vibration? Its amplitude would be so small that a very sensitive detector would be needed. Weber decided that a piezoelectric detector would work best. Piezoelectric crystals have the property of generating a small voltage when squeezed, and they would be capable of detecting a slight change in the aluminum cylinder as the gravitational wave passed. He glued several of them around his antenna.

The aluminum cylinder was five feet long by just over two feet wide. It weighed 2,600 pounds and was suspended by wires so that it was free from disturbances. Considerable shielding was needed to isolate it from

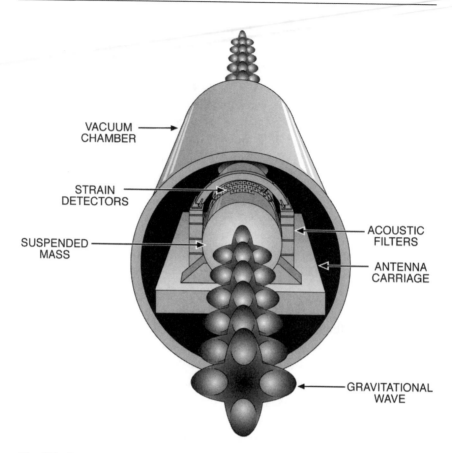

VACUUM CHAMBER

STRAIN DETECTORS

SUSPENDED MASS

ACOUSTIC FILTERS

ANTENNA CARRIAGE

GRAVITATIONAL WAVE

Fig. 35: Cutaway showing Weber's gravitational wave detector. A gravitational wave is shown striking it.

the numerous disturbances that would vibrate it. The resonant frequency of the antenna was 1,661 Hz (vibrations per second), which corresponded to a wavelength of about 110 miles. This was near the expected frequency of gravitational waves.

Weber put his antenna into operation in 1963, and over the next three years, he recorded large number of pulses. But Weber was worried that the pulses might not be a result of gravitational waves. He needed something more and therefore decided to built a second antenna at Argonne National Laboratory near Chicago, about seven hundred miles away. A gravitational wave passing through the first detector would strike the second a fraction of a second later. Weber looked only for coincident events of this type, and in December 1968 both detectors recorded events at approximately the same time. Over the next several months more than a dozen such events were recorded.

Weber was delighted, but to be on the safe side, he and his team checked everything to make sure their results weren't spurious. He finally convinced himself that they were valid, and in June 1969 he announced that he had detected gravitational waves. Scientists were amazed, and soon the popular press picked up on the discovery and a lot of hoopla followed. If true, it was a monumental discovery. Within a short time Weber announced that he believed the waves were coming from the center of our galaxy.[4]

OTHERS LOOK FOR THE WAVES

Weber's announcement caused so much excitement that other groups soon began setting up equipment to look for the waves. One of the first was Vladimir Braginsky of Moscow State University. He constructed two cylindrical antennas similar to Weber's, but he used a slightly different detector. Over several months he detected a number of events, but he eventually decided that they were spurious.

J. A. Tyson of Bell Laboratories in New Jersey also constructed an antenna similar to Weber's, but tuned to a different frequency (710 HZ), and after several months of operation, he reported that he had also failed to detect anything. Ronald Drever of the University of Glasgow in Scotland became excited about gravitational waves after attending one of Weber's lectures. He immediately began working on a detector, but like Braginsky and Tyson, he found no evidence of the waves. Finally, Richard Garwin and James Levine of IBM built a detector and found nothing.[5]

Robert Forward and Gaylord Moss of Hughes Research Laboratories used a different method to look for the waves. They used an almost-free device with two suspended masses. Their detector was broadband, so it could detect waves over a range of frequencies. But as with the other experimenters, no waves were detected. By now most scientists were becoming skeptical of Weber's discovery. Yet strangely, while everyone else was finding nothing, Weber continued to report new events. In fact, the frequency and intensity of his events led to even more skepticism.

Theoretical calculations showed that he was getting many more events than could reasonably be expected.

SOURCES OF GRAVITATIONAL WAVES

What could be causing the events that Weber was seeing, if they were, indeed, valid events? We saw earlier that Einstein showed that a binary star system would be a relatively weak source. But it has one of the major requirements: asymmetry. Gravitational waves are only given off by asymmetric sources. A spherical pulsating star, for example, will not produce gravitational waves, and neither will a spherical spinning star. A binary system, on the other hand, will, since it is not symmetric; in other words, it has the required asymmetry. But the components of the system have to be exceedingly dense and close to one another. A good source would be two neutron stars whirling around one another or colliding with one another. Freeman Dyson of the Institute for Advanced Study at Princeton, New Jersey, made an estimate of the radiation emitted when two neutron

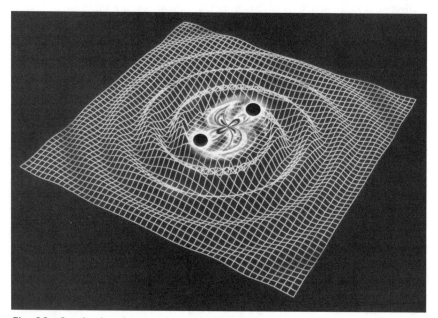

Fig. 36: Gravitational waves being given off by
two black holes revolving around one another.

Fig. 37: Gravitational waves given off in a supernova explosion.

stars collided. His calculations seemed to indicate that Weber's equipment was capable of detecting the gravitational waves that would be produced.

Even better than two neutron stars, however, would be two black holes in a binary system (figure 36).[6] They would give off gravitational waves of much greater intensity. Furthermore, as they lost their energy, the two black holes would draw closer together and revolve around one another faster, which would cause them to give off even more energy. Finally, they would fall into one another with a tremendous release of energy. How many collisions of this type would we expect in a year? Theorists are still uncertain, but a large number of calculations are in progress using supercomputers. And the general consensus is that the number is small, with estimates being in the neighborhood of one or two.

Another possible source of gravitational waves is supernovae. The explosion would have to be asymmetric, but most supernovae remnants show signs of asymmetry. The remnant of a supernova is a neutron star, which by itself would not be a good source. However, there is some evidence that when they form, neutron stars may undergo distortions before settling down. These distortions would not last long, but they would be an excellent source of waves. Furthermore, even after a neutron star settled down, it might develop small irregularities on its surface. Small mountains, about an inch high, if frozen in place, would serve as a good source as the star spun on its axis.

Radio galaxies and quasars, and even ordinary galaxies, are now believed to have gigantic black holes at their core. These holes may be devouring stars in their neighborhood. If so, they are another potential source of gravitational waves.[7] Furthermore, gravitational waves were likely generated shortly after the universe was created, in the big bang explosion. If so, these waves might form a background similar to the cosmic background radiation. If we could detect this radiation, it would tell us a lot about the early universe, since it would go back to 10^{-43} second after the big bang. No other source is this old.

Theoreticians also tell us that several "defects" may have been generated in the big bang. Two of these are cosmic strings and magnetic monopoles. Cosmic strings would be extremely long and tremendously massive; calculations indicate that a section only an inch long could weigh as much as Earth. As they whip through space, they would generate gravitational waves. Monopoles, or particles with a single magnetic pole, may also exist, and they would also be potential sources of gravitational waves, since they would be exceedingly massive.[8]

Finally, there are, of course, things that we haven't even imagined. We know that science in a hundred years, or even in fifty years, will be quite different from that of today, and many new discoveries will no doubt be made.

TAYLOR AND HULSE

As the years passed, and more and more groups detected nothing, scientists began to accept the inevitable: gravitational waves had not been discovered. Furthermore, calculations were now showing that they were going to be much more difficult to detect directly than originally anticipated. Pessimism soon began to spread. But there was another way: it might be possible to detect them indirectly. And, indeed, this is what happened.

Joseph Taylor received his doctorate in radio astronomy from Harvard University in 1967. The following year he accepted a postdoctoral at Harvard. Pulsars (short-period radio stars) had just been discovered, and Taylor was particularly interested in finding new ones, so, along with several colleagues, he wrote a computer program that could be used in conjunction with a radio telescope to help in the search. And within a few months he had discovered several.

In the fall of 1969 he joined the faculty of the University of Massachusetts at Amherst, where he hoped to continue his search. By now several dozen pulsars had been discovered, and Taylor was eager to use his computer program, which had now been improved. He hoped that his program would help him find a pulsar in a binary system.

In 1970 graduate student Russell Hulse approached Taylor, asking if he could work on his pulsar team. Taylor was eager to oblige him and assigned him a project using the huge one-thousand-foot radio telescope at Arecibo in Puerto Rico. A pulsar had recently been discovered in the Crab Nebula that pulsed at the incredibly fast rate of thirty pulses per second, and Taylor was interested in finding other fast pulsars: he assigned the project to Hulse. Hulse went to Arecibo in December 1993, and within a short time he found a source on the border between the constellations Sagitta and Aquila that appeared to be pulsating very rapidly. It was referred to as PSR 1913+16 (the numbers refer to its coordinates). He finally determined that it was pulsing at the rate of seventeen pulses per second, making it the second fastest pulsar on record.

But the source puzzled Hulse. He measured its rate several times, and each time he got a slightly different result. Thinking his instrumentation was at fault, he checked everything carefully, but nothing seemed to be wrong. Finally, he realized the pulsar's period was, indeed, changing slightly, but in a regular way. This is what a binary system would show. Hulse was excited. He immediately telephoned Taylor and told him the news.

Within a short time they had determined the mass of the pulsar was 1.4411 solar masses, which was the correct range for a pulsar, or a neutron star, as pulsars had now been shown to be. There was some difficulty in determining the mass of the second component of the system, but it was eventually shown to have a mass of 1.3873 solar masses. The two stars were orbiting one another at a distance of 435,000 miles (slightly larger than the diameter of the Sun).

Hulse finished off his thesis and went on to other projects, but Taylor continued to study the system. He soon realized that it could be used as a check on the general relativistic gravitational wave formula. Two neutron stars revolving around one another would be a good source of gravitational waves. But both electromagnetic and gravitational waves would be given off, and most of the energy released would be in the form of electromagnetic waves, so it wouldn't be an easy task.

As the system gave off energy, the two stars would get closer and closer together, and as a result, the period of the system would decrease. Calculations were made and it was found that the period should decrease by 1/10,000 (10^{-4}) of a second per year. This was obviously an extremely small amount and would be difficult to measure. But Taylor persevered; for four years he and his colleagues, Peter McCulloch of the University of Tasmania and graduate student Lee Fowler of the University of Massachusetts, monitored the system. By 1978 they had made over a thousand observations on it and had determined that over the four years its period had decreased by .000414 second. This was in agreement with general relativity, and in 1978 Taylor made the announcement at the Ninth Texas Symposium on Relativity in Munich, Germany. He published the results in *Nature*. The announcement generated a lot of excitement, but many people were skeptical at first. Within a few months, however, most scientists were convinced. Gravitational waves had been detected at last—not directly, but indirectly. Hulse and Taylor received the Nobel Prize for their discovery in 1993.[9]

WEISS AND DREVER

Two men who would play an important role in gravitational wave astronomy over the next few years were Rainer Weiss and Ronald Drever. Weiss was born in Berlin in 1932 and came to the United States with his family in 1939; they settled in Manhattan. As a youth he liked mechanical things and electronics, but when he went to MIT, he majored in physics. As a graduate student he became fascinated with general relativity. After receiving his doctorate, he was taken on as an assistant professor at MIT, and while teaching a class in general relativity, he became interested in gravitational waves and decided to look into the design of a detector. He wanted a new type of detector, something different from

Weber's bar detector, and he came up with one based on interferometry (the interference of two superimposed waves) that appeared to have considerable promise. Robert Forward of Hughes Laboratories heard of his design, and he and Gaylord Mass and Larry Miller built a prototype. After three years of work, they completed it in December 1972, but like the others, they found no evidence of gravitational waves.

Ronald Drever obtained his doctorate in nuclear physics in the mid-1950s. Throughout the 1960s he worked in nuclear physics and cosmic ray physics, but one day during a stay at Oxford, he went to a lecture by Joseph Weber on gravitational waves. Drever was intrigued and immediately decided to build a detector. He knew he didn't have the resources to build a resonant bar detector that was better than Weber's, so he decided to build an interferometer detector. He had considerable experience in interferometry and was convinced it would be the least expensive way to go. As it turned out, it was a much more difficult and expensive task than he had anticipated, but he was eventually able to build the detector. But, like the others, he found nothing.

THE INTERFEROMETER DETECTOR

Let's consider how the interferometer detector differs from the resonant bar detector. We'll begin by considering the basic principle of interferometry. In an interferometer two beams of light are superimposed so that they interfere with one another. If the two beams are exactly in phase so that their peaks and troughs line up, the light is enhanced. If the two beams interfere out of phase so that the troughs of one line up with the peaks of the other, the light is cancelled, and there is darkness. In between these two extremes, some light will be visible, but it will be less than the maximum.[10]

Weiss and Drever used laser light in their interferometers. Their systems were in the shape of an L, with arms of equal length. Mirrors were placed along each of the arms. A laser beam entered the system at the corner of the L and was split into two beams, with one beam directed down one arm and the other down the other arm. Mirrors were placed near the center so that the beams could be bounced back and forth many times. Eventually, they were recombined and interfered with one another.

When a gravitational wave passed through the system, one of the arms would contract and the other would expand, and the movement (as slight as it was) could be seen in the signal from the detector. One of the main advantages of this type of detector over the resonant bar detector was that the longer the arms, the more sensitive the detector, and the arms could easily be made relatively long. Furthermore, the number of reflections between the mirrors also determined the sensitivity, and they could easily be increased. Finally, the interference detector was broadband in that it could detect waves of many different frequencies. In contrast, the bar antenna was tuned to only one frequency. Because of these advantages, almost all of the detectors in recent years have been interference detectors.

KIP THORNE AND THE INITIATION OF LIGO

After several years of working in black hole physics, Kip Thorne of Caltech wanted a change.[11] A large number of scientists were now working in black hole physics, and he decided to search for a more fertile area. He had become interested in gravitational waves, particularly after Weber's announcement, and he felt there was tremendous potential in the field, since it was still in its infancy. Most of his work was theoretical, but he realized it would be advantageous to get an experimental program going at Caltech. He knew that a lot of money would be needed, and the project would be risky because gravitational waves were exceedingly weak and therefore difficult to detect. Nevertheless, in 1976 he approached the Caltech administration and the National Science Foundation (NSF) with a proposal for a large gravitational wave detector. The administration gave him their approval on the condition that he came up with an outstanding experimentalist to direct the project. Thorne's first thoughts were of Braginsky of the University of Moscow, but Braginsky turned him down. He was reluctant to leave Moscow—worried that the KGB would take it out on his family if he left (the Cold War was still on at this stage). Thorne then turned to Russell Drever in Glasgow. Drever was hesitant at first, but eventually agreed.

Drever, who had worked with interference detectors in Glasgow, wanted to continue using them at Caltech. Thorne was hesitant at first; at this point he still preferred resonant bar detectors. But he was soon convinced that the interferometer detector was superior. The initial aim of the project was to build a prototype interferometer detector at Caltech, then

Fig. 38: Kip Thorne.

build larger models at two sites to be selected that would be used in the search for gravitational waves. During the early 1980s Drever and his team built a forty-meter prototype to make sure that there were no flaws in the design.

While Drever and his team were working on their detector, another team under Rainer Weiss was working on a similar design at MIT, and both were being funded by the NSF. In a review of the projects in 1984, the NSF suggested that the two teams merge. Both directors were stunned, and resisted strongly, but in the end they had no choice. A new project was formed with Drever and Weiss as codirectors; Thorne was also still actively involved in the project. It was, however, a rocky marriage, since Weiss and Drever did not get along well, and in 1986 a committee reviewing the project suggested that a single director would be more appropriate. Robbie Vogt, who had worked at NASA's Jet Propulsion Laboratory, was brought in. Barry Barish later replaced him.

LIGO

Once the prototype was perfected, two sites were selected, one at the Hanford Nuclear Reservation in south-central Washington State and the other near Livingston, Louisiana.[12] The sites were referred to as the Laser Interferometer Gravitational Wave Observatory (LIGO); each consisted of an L-shaped complex, and they were separated by 1,900 miles. Both sites are remote, so as to minimize outside disturbances. Two facilities are needed for the same reason that Weber used two detectors. Both detectors are subject to many different sources of "noise," and to be sure that the signal is real, it must happen simultaneously at the two detectors. Gravitational waves would trigger both detectors at about the same time; spurious sources would not.

Both detectors have arms that are 4 kilometers (about 2.5 miles) long. Actually, there is a second detector at Hanford that has 2 kilometer arms. Detectors at both facilities can receive frequencies from 100 to about 3,000 Hz, which is within the expected range for gravitational waves. The end mirrors are 4 kilometers from the laser, and the beams pass along evacuated tubes. The mirrors are made of fused silica that is exceedingly pure and uniform; each weighs twenty-two pounds. The surfaces are ground to incredible precision, and the laser beam strikes each of them 130 times. The mirrors are hung from fine steel wires, a mere hundredth of an inch in diameter. They are free to move when a gravitational waves passes them.

Critical to the success of the observatory is isolating the detectors

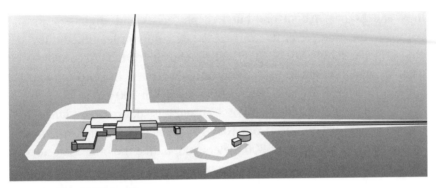

Fig. 39: The LIGO facility.

from external noise and disturbances. Seismic tremors, tides from the Sun and Moon, passing trucks, jets flying overhead, and lightning are only a few of the things that give rise to disturbances. They must all be filtered out, and a considerable amount of work has been done to make sure this is the case.

Observations began in May 2002. The hope is that the detectors will eventually receive gravitational waves from neutron binary systems, black hole binary systems, and supernovas, and possibly from other sources. The sensitivity is three parts in 10^{23} at 180 Hz, which is incredible sensitivity (millions of times smaller than a nucleus). Most astronomers in the field are convinced that gravitational wave astronomy will eventually become a major branch of astronomy, since signals of this type and age cannot be received from any other instrument.

GEO 600, VIRGO, AND TAMA 300

LIGO may be the biggest project of its kind in the world, but it is not the only one. Several other projects are either already on line or scheduled to start soon. GEO 600 is one of them; it is located in Germany, near Hanover, and is a joint project of Germany and Scotland. The arm lengths are 600 meters, and it has sensitivity of eight parts in 10^{21} at 600 Hz. GEO 600 went into operation in 2002.

Another project, referred to as VIRGO, is a collaboration between

Italy and France. About one hundred physicists and engineers are involved in the enterprise. Its arms are 3 kilometers in length, so they are only slightly shorter than LIGO's. It is named for the Virgo cluster of galaxies, which might be a good source of the waves. Its sensitivity is one part in 10^{22} at 500 Hz, and observations started in 2003.

A third project is called TAMA 300, and it is sponsored by Japan. It is located about twelve miles from Tokyo at Japan's National Astronomical Observatory. It has arms of 300 meters and, unlike the other detectors, it is completely underground in long concrete tunnels. It has a sensitivity of 5 parts in 10^{21} from 700 to 1000 Hz. Observations began in 2001.

Australia also has plans for a gravitational wave observatory. It is to be built near Perth. The above observatories, together with LIGO, will act as a worldwide network and will reinforce one another and serve as checks on one another.[13]

NEXT-GENERATION DETECTORS

As you might expect, plans to upgrade the present detectors are already on the drawing boards. Beginning in 2006, LIGO will be boosted in power from 10 to 180 watts, and the steel wires that hold the mirrors in place will be replaced by silicon ribbons. Furthermore, the silica glass mirrors will be replaced by sapphire crystals. This will boost the sensitivity of the system by a factor of about twenty. The new system will be referred to as LIGO II.

The Japanese are also planning on a successor to TAMA 300. The new detector will have 3-kilometer (slightly under 2 miles) arms, and will again be underground; it will use sapphire mirrors. The most ambitious project of the future, however, is a joint project between NASA and the European Space Agency. It is a space gravitational wave observatory that they have named LISA. The interferometer arms are to be 5 million kilometers (3 million miles) long. At the ends of the arms will be satellites; again, the overall form is an L. They will orbit the Sun. This project will be even more expensive than LIGO, which cost $365 million, so there's no guarantee it will go forward in the near future, but plans are being made nevertheless.[14]

EINSTEIN'S LEGACY

It all started with the prediction of gravitational waves by Einstein and the determination of a formula for predicting their intensity for various sources. Einstein was sure they would never be detected and no doubt would be surprised at the worldwide effort now going into detecting them. Hundreds, and perhaps thousands, of people in the world are now participating in one way or another in the search for the waves. It is another of Einstein's many legacies.

Chapter 6

Gravity's Cosmic Lenses

In 1911 Einstein showed that light would be deflected by a massive object, and a few years later he used it to formulate his general theory of relativity. Einstein concentrated on the deflection caused by the Sun, since it was the most massive nearby object. He calculated how much a beam of light from a star would be deflected as it grazed the Sun—a prediction that was verified in 1919 during an eclipse expedition by Arthur Eddington of Cambridge University.

But if the light beam from a star is deflected by the Sun, it will also be deflected by any massive object in space. In particular, if a star is behind a massive object, its light rays will be bent as they pass the object. In effect, the object will act as a lens. Someone wrote to Einstein in 1936, asking him if this was possible, and Einstein immediately looked at the problem.[1] Working out the details, he found that a massive object would, indeed, magnify distant objects such as stars. But when he looked into the probability that such an event would actually occur, he concluded that it was unlikely. The two objects would have to be exactly aligned, and Einstein was sure the probability of this was small. Nevertheless, he published an article in *Science* discussing the possibility and pointing out its consequences. We now refer to it as a gravitational lens.[2]

Few people paid any attention to Einstein's paper. Most astronomers agreed that it was a long shot observationally. But one astronomer took it seriously. Fritz Zwicky of Mt. Wilson Observatory in California became intrigued with the idea. He realized that the probability of two stars being aligned was exceedingly small, but he found that the alignment of two galaxies was likely. Unfortunately, he didn't have the equipment to check on his idea, and no one followed up on it. Henry Norris Russell of Harvard University also noticed Einstein's article and wrote a paper suggesting that gravitational lenses could be used to "visualize and popularize" relativity. It was something people could easily understand.

STRUCTURE OF A GRAVITATIONAL LENS

What are the major components of a gravitational lens? There are three: a background object, a deflector, and an observer on Earth. Quasars are excellent candidates as background objects because they are the most distant objects in the universe. Because they are so far away, they are more likely to have some sort of object in their line of sight. Distant galaxies are also good candidates.

The deflector, or lensing object, could be a galaxy or even a black hole. Black holes would make particularly good deflectors because they are so compact and massive. If the deflector and the background object are exactly aligned with Earth, an observer will see a ring, which is, of course, an enlarged image of the quasar; and superimposed at the center of it will be the deflector. The deflector magnifies the distant object, but because it isn't a perfect lens, we won't see the central part of the quasar. To understand why, let's compare a gravitational lens to an ordinary glass lens.[3] In a glass lens, the rays from all parts of the lens focus to a single point, and we see an enlarged image. In a gravitational lens, the rays do not focus to a single point; they focus to a line behind the lens, and as a result we don't see a focused object.

Very few objects in the universe are in perfect alignment with Earth. A more likely case is two objects that are closely aligned, but slightly off. In this case an observer on Earth would see two or more images. If we could see them close-up, they would look like tiny crescents, but from a distance they would appear as two or more points of light. If we examine these points, however, we will find that they are identical. In particular, their spectra will be identical, and this is because they are, in reality, the same object.

Another difference between gravitational lenses and glass lenses is that glass lenses bend different frequencies of light by a different amount. We see this effect in the prism, with the spreading of light into a rainbow of colors. The gravitational lens, on the other hand, bends all wavelengths by the same amount. Indeed, it bends x-rays in the same way it bends ordinary light.

EARLY PREDICTIONS

For many years after Einstein made his prediction, there was little interest in gravitational lenses. They seemed to be far-fetched, without any practical applications, and astronomers had never seen any evidence of them. Nevertheless, it was an intriguing phenomenon, and theorists eventually began to take notice. Sjur Refsdal of the University of Hamburg worked out the mathematics for several applications of the phenomenon and showed that it could become an important tool in astronomy. In particular, he pointed out that it could be used to determine the constant of cosmology called Hubble's constant.[4] Hubble's constant is the recessional velocity of a galaxy divided by its distance. Because it is related to the age and size of the universe, it is of fundamental importance to cosmologists. For a while other theorists followed up on Refsdal's work, but eventually interest died.

Then in the mid-1960s quasars were discovered. They were strange objects in the outermost reaches of the universe that appeared to be "overbright," in other words, too bright for their distance. No one was quite sure what they were at first. Jeno Barnothy suggested that they might be two galaxies in alignment, so that the light of the background galaxy was enhanced. This would explain their excessive brightness. But as more and more quasars were discovered, the explanation seemed less likely. Indeed, thousands are now known, and we know that the explanation is incorrect.

Fig. 40: The Multiple Mirror Telescope in Arizona.
(Photo courtesy of MMT Observatory.)

DISCOVERY OF THE FIRST GRAVITATIONAL LENS

Although the discovery of quasars renewed interest in gravitational lensing, most astronomers were still skeptical. Then in May 1979 came an announcement that took them by surprise. Dennis Walsh of the Jodrell Bank Radio Telescope Observatory, Robert Carswell of Cambridge University, and Ray Weymann of the University of Arizona announced in *Nature* that they had detected what they believed was a gravitational lens.[5] The quasar 0957+561 (the numbers refer to the coordinates of the quasar) consisted of two components that appeared to be identical. Separated by a mere six seconds of arc in the sky, they had the same redshift and their spectra were identical.

Was it possible that they were identical twins, or were we seeing a double image? The three astronomers checked everything carefully and, indeed, they were virtually identical. It was almost too much of a coincidence. Walsh, Carswell, and Weymann decided that it had to be a gravitational lens. But there were difficulties. There were two types of lines in the spectra: dark lines and bright lines. Dark lines are usually found in galaxies and stars, but bright lines were seen only in gaseous nebulae. The bright lines seemed to indicate that the quasar was surrounded by an expanding envelope of gas.

It soon became clear that higher-quality spectra were needed. As it turned out, a new, more powerful telescope had just been completed on Mt. Hopkins in Arizona. Called the Multiple Mirror Telescope (MMT), it consisted of six 1.8 meter mirrors, which, through the use of computers and lasers, were synchronized to work as a single unit (figure 40). With the MMT, new spectra was obtained, and as hoped, it was of much higher quality. It verified that the two objects were identical. But there was a problem: if it was a gravitational lens, there had to be a deflector somewhere between Earth and the quasar, and there was no sign of one.

The object then moved into the day sky, and astronomers could do little but wait for it to return to the night sky. Radio astronomers, however, do not need a dark sky, and the radio astronomers at the Very Large Array (VLA) in Socorro, New Mexico, trained their radio telescopes on it. To their surprise, they found that the object was much more complex than it originally seemed. In addition to the two main images, there were three other faint images that appeared to be part of the system. But again there was no sign of the deflector. Then Alan Stockton of the University of Hawaii, using a telescope on Mauna Kea, obtained an image that showed that one of the two main images had a faint fuzziness on one edge. Astronomers quickly jumped on the find. A group at Palomar Observatory, consisting of Peter Young, James Gunn, Jerome Kristina, Beverley Oke, and James Westphal, used a charged coupled device (CCD) to examine the fuzziness. Looking it over carefully, they realized it was what everyone was searching for. Superimposed on one of the quasar images was a foreground galaxy: the deflector. This was quite a surprise; astronomers hadn't expected it to be coincident with one of the images. Indeed, further examination showed that the galaxy was actually part of a cluster. This explained the other images seen in earlier photographs.

By the mid-1980s astronomers were confident that the first gravitational lens had, indeed, been discovered. Then a second one was discovered. Ray Weymann and several colleagues discovered one with three images. And within a short time several others were discovered.

GRAVITATIONAL LENS SURVEYS

The discovery of several quasar gravitational lens systems soon sparked an interest in large-scale searches. In these searches astronomers begin by using radio telescopes to look at large numbers of quasars and galaxies, searching for any that appear to have multiple images. Once good candidates are found, they study them in detail using optical telescopes. One of the first surveys of this type was initiated by Edwin Turner of Princeton University and several colleagues.[6] Turner stated that the rate of discovery was low at first, with only one or two being found a year. Later, however, the rate increased. Tony Tyson of Bell Labs, who was involved in the search for gravitational waves, also initiated a gravitational lens survey.

In recent years the most comprehensive gravitational lens survey was called CLASS (Cosmic Lens All Sky Survey).[7] About ten thousand radio sources were checked, with seventeen gravitational lens systems being found. Today, astronomers know of about sixty double, triple, or multiple quasars that appear to be gravitational lenses. Overall, compared to the number of known quasars, they are relatively rare, with only about one in five hundred of all observed quasars in a lensing system. Many uncertainties still exist when it comes to identifying a quasar lensing system, so we are still not sure how many there actually are.

DETERMINATION OF THE HUBBLE CONSTANT

As we saw earlier, Hubble's constant is one of the most important constants in cosmology. It gives us information about the age and size of the universe. The usual procedure for determining it is complicated, involving a step-by-step ladder to the outer reaches of the universe, and if any of the rungs of the ladder are faulty, the entire estimate is thrown off. Sjur Refsdal of Germany showed that lensing could also be used to determine the Hubble constant, and a cosmic ladder would not be needed; the constant could be calculated directly.[8]

The key to the calculation is the slight difference in time that it takes for light signals from the two images to reach us. This difference is a result of the slightly different paths the two light rays take and the gravi-

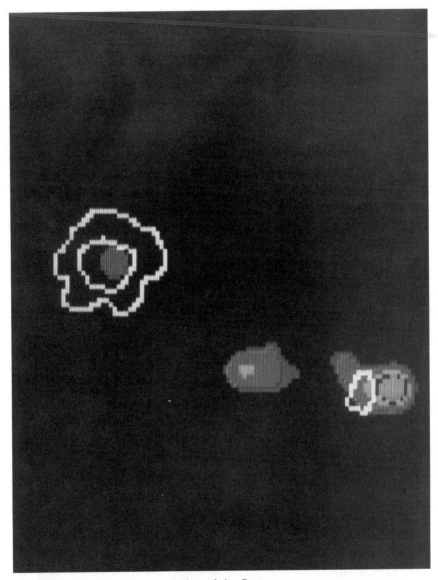

Fig. 41: A computer representation of the first
gravitational lens showing the positions of the images.
(Photo courtesy of the National Optical Astronomy Observatories.)

tational field of the lensing object. Using a model of the mass distribution and shape of the lens, astronomers can use the time delay to determine the distance to the quasar. Its recessional velocity is given by its redshift, which we get from its spectrum. Hubble's constant is the ratio of these two numbers. The technique was applied to 0957+561 with considerable success. Since that time, it has been applied to several other similar systems. The values obtained are slightly lower than obtained by the usual methods, but it is a good check.

THE COSMOLOGICAL CONSTANT

Another of the constants of cosmology is Einstein's cosmological constant. He used it to stop the expansion of the universe in his original theory, but in 1932, after it was discovered observationally that the universe was expanding, Einstein discarded it. Others, however, kept it, and for many years it continued to be a source of controversy. Then, with the discovery of the acceleration of the expansion of the universe, it was again taken seriously. Its value cannot be calculated directly from gravitational lensing, but the phenomenon gives considerable insight into it. The reason is that the acceleration of the universe, which may be directly related to the constant, increases the number of quasars that are lensed. This means there is a link between the number of observed quasar lenses and the constant. Using this link, Emilio Falco, Chris Kochanek, and Jose Munoz of the Harvard-Smithsonian Center for Astrophysics made an estimate of the cosmological constant and showed that it could account for up to 62 percent of the energy density of the universe. It is expected that gravitational lensing will provide more information about the constant in the future.[9]

EINSTEIN RINGS

One of the most spectacular lensing effects is the Einstein ring. It is a complete circle of light with the deflector at the center. As we saw, this requires a perfect alignment of the background object and the deflector, so that the rays from the background object bend around the deflector symmetrically. The chances of this are quite low, so only a few Einstein rings are known. The first was discovered in 1988 by J. N. Hewitt, Edwin Turner, D. P. Schneider, B. F. Burke, and G. I. Langston. It was identified with the radio source 1131+0456 and had a diameter of about 1.75 arc seconds. A particularly spectacular one was discovered by L. J. King and several colleagues. It exhibited an almost perfectly circular ring plus a bright central galaxy. It is associated with radio source 1938+666.

In all, about a dozen Einstein rings are now known.[10] They have diameters ranging from .33 to about 2 arc seconds. Some of them, however, are not complete rings, but close enough to qualify as Einstein rings. As with multiple-image lenses, Einstein rings can also be used to determine the Hubble constant and to identify dark matter. They also help us in understanding the mass distribution of galaxies.

GIANT ARCS AND ARCLETS

In some cases the lens is not a single galaxy, but a cluster of galaxies. In this case we usually get a large number of arcs; some of them are large, but most are small. We refer to the smaller ones as *arclets*. The first of this type was discovered by Roger Lynds of the National Optical Astronomical Observatory and Vahé Petrosian of Stanford University. It was also discovered independently by Genevieve Soucail of Midi-Pyrénées Observatory in France and her colleagues. Almost one hundred such arcs clusters have now been identified.

The giant arcs are helpful in two ways. First, they are galaxies that we would not be able to see if they were not in a gravitational lens, and many of them are in early evolutionary stages, giving us information about the formation of galaxies. The second way they are useful is that we can use them to study the mass distribution of such systems. Systems such as these are also helpful in our study of dark matter. Most galaxy clusters are dominated by dark matter, and a study of the lensing in the system gives us information about the dark matter in the cluster.

One of the most spectacular systems of this type is Abell 2218. It contains several giant arcs and more than one hundred smaller arclets. Several such systems have recently been found with the Hubble telescope.[11]

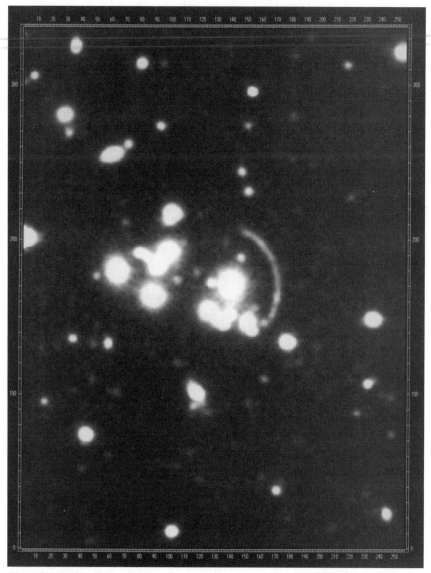

Fig. 42: An Einstein arc.
(Photo courtesy of the National Optical Astronomy Observatories.)

MICROLENSING OF QUASARS AND STARS

In some cases the lensing of quasars is quite subtle. For example, if a star passed in front of a quasar, as it got close to perfect alignment, the quasar would break up into two or more images; then, as perfect alignment occurred, a ring would appear. But in most cases the images are so faint and close together that they cannot be resolved with present-day telescopes. Nevertheless, there is a slight change in brightness that can be detected. As the star nears alignment, the quasar brightens slightly, then, as it passes the quasar, it dims. Unfortunately, there's a problem. Quasars are variable, and their brightness is continually changing. We have to be able to distinguish this internal change in brightness from the change due to lensing. Astronomers have studied the changes in brightness, and, as it turns out, there is a difference in the two cases. If the change is internal, both images (or three or more, whatever the case) will change in the same way. On the other hand, if a star passes in front of the quasar, it will eclipse one image at a time. This means that during lensing only one image will change. Astronomers have now studied about a half dozen lensings of this type. They are referred to as microlenses.[12]

Microlensing can also be used to detect dark objects such as dead stars and black holes. If a black hole, for example, passes in front of a star, it will create several images and an overall brightening of the star. Again, it's unlikely that astronomers could resolve the images, but they could measure the brightening. The chances of such a lensing occurring at any time are extremely small—only about one in a million. Astronomers therefore monitor millions of stars, looking for tiny changes in their brightness.

Several teams have looked into this method of detecting dark objects; they include a French team called EROS, an American-Australian team called MACHO, and a Polish-American team called OGLE. Over a period of seven years, the teams have found two dozen microlensings in the large Magellanic Cloud, a nearby galaxy.[13] Over five hundred microlensings have been detected in the direction of the center of our galaxy, the Milky Way. Some of these, however, are no doubt eclipsing binaries, but a few may be caused by black holes.

EXTRASOLAR PLANETS

One of the most exciting astronomical events that has occurred in recent years is the discovery of planets orbiting distant suns. So far, we haven't detected any the size of Earth; nevertheless, we now have evidence that planets do exist in space beyond our solar system. The usual methods for searching for these planets are difficult to perform and tedious to carry out, and so they require a lot of time. Furthermore, they can detect only very large planets, of the size of Jupiter. Earth-sized planets are beyond their range. Lensing may overcome this. Indeed, it should be capable of detecting Earth-sized planets.[14]

As in the above cases, the idea is to search stars for a sudden burst of brightness. This is what would happen if a planet passed across the face of the star; it would last for only a few hours and increase the brightness by only a few percent, but it would be detectable. Several teams are presently involved in searches of this type. One, called PLANET (Probing Lensing Anomalies Network), is led by Penny Sackett of the University of Groningen in the Netherlands. Another, called MPS (Microlensing Planet Search), is headed by David Bennett of the Univeristy of Notre Dame. And, finally, a third group, called MOA (Microlensing Observations in Astrophsyics), is led by Philip Yock of Aukland University in New Zealand. Several good candidates have been found. The Hubble space telescope has also got into the act and has detected several good candidates.[15]

COSMIC STRINGS

Earlier we saw that defects called cosmic strings may have been left over from the creation of the universe. They would be very thin, but incredibly massive, with an inch weighing as much as Earth. With this much mass, they would be capable of lensing. No one gave the idea much thought until Len Cowie and Esther Hu of the Institute of Astronomy in Hawaii found a line of three identical pairs in the sky. It was too much of a coincidence. The only explanation seemed to be that a cosmic string had created them, and Cowie and Hu published a paper making the suggestion. It may be our first direct evidence of the existence of cosmic strings.[16]

THE FUTURE

The future of gravitational lensing looks bright. It has already become a major tool in astronomy, and it is quite likely to become even more important in the future. And, again, Einstein was at the forefront of the discovery. He worked out the details of lensing and predicted that distant objects would be magnified if there was an exact alignment in the sky. Furthermore, he even predicted that several images would be seen if the alignment was slightly off. He realized, however, that the chances of such an alignment were small, and he thought the effect would end up being nothing more than a scientific curiosity. As we have seen, he was wrong. Lensing has become an important tool. It has proven to be of considerable value in cosmology, in detecting the structure of quasars and galaxies, and in the search for dark matter. Furthermore, it may help us find Earth-sized planets in the near future.

Chapter 7

Einstein's

Quantum Legacy

E instein is best known for his special and general theories of relativity, and they are generally considered to be his greatest contributions to physics, but his contributions to quantum theory were also extensive. He spent a considerable amount of time thinking about quantum theory and its implications. According to his friend Otto Stern, he once said, "I have thought a hundred times as much about the quantum problems as I have about general relativity."[1] And there is no doubt that he had considerable admiration for the theory, but he struggled with its implications and could never fully accept it. It was a theory based on chance and probability, and he had been brought up on classical theories in which everything could be predicted exactly. In quantum mechanics you could determine only probabilities; it was indeterminate, and this troubled Einstein. In 1926 he said to Max Born, "Quantum mechanics is very worthy of regard. But an inner voice tells me that this is not yet the right track."[2]

The difficulty that Einstein had in accepting quantum theory might seem a little strange in that he made several important contributions to the theory in its early stages. But after quantum mechanics was formulated, Einstein acted mainly as a critic. His criticisms, however, were not trite; they played an important role in giving scientists a deeper understanding of the theory, and he pointed out some serious difficulties with it. One of his most famous papers on quantum theory, published in 1935, is now considered by many to be one of the most important papers ever published in physics. It shocked the scientific world and caused several scientists considerable grief, but in the end it gave us a greater understanding of the theory.[3]

Quantum theory was born in 1900 when Max Planck of Germany found a solution to a problem that had been bothering physicists for years. They could not explain the emission of radiation from a heated object. Planck tried a radically different approach using the idea that tiny "oscillators" emitted the radiation in tiny chunks. He referred to these chunks as *quanta*. Even though his idea explained the phenomenon beautifully, few took it seriously. It was too radical. Einstein, who was working in a patent office in Bern at the time, looked carefully at Planck's approach and realized it was a significant breakthrough. He extended the idea to radiation itself, stating that light and other forms of radiation were made up of particles, later called *photons*. In short, it was both a particle and a wave, and he used the particle point of view to explain a phenomenon that had been discovered a few years earlier, called the *photoelectric effect*. Philipp Lenard of Germany had shown that electrons are emitted from the surface of a metal when light strikes its surface. But only certain frequencies produced electrons, and their energy did not depend on the intensity of the light; it depended only on its frequency. Einstein assumed the light was composed of photons that struck the electrons in the metal and knocked them free. A certain minimum energy, however, was required to free them. Many years later, in 1922, Einstein received the Nobel Prize for this work.[4]

In 1913 Niels Bohr of Denmark used quantum theory to formulate his model of the atom. Scientists soon realized that quantum theory was an important new tool, which was needed when dealing with the microworld. But for many years the theory was a hodgepodge of seemingly unrelated ideas. Then in 1925 Werner Heisenberg of Germany brought things together in a theory that we now refer to as *quantum mechanics*. It was a strange theory, however, based on arrays of numbers called matrices, and most scientists were not familiar with these arrays, so, at first, there was considerable reluctance to accept it. Soon after it appeared, however, another quantum mechanical theory was formulated by Erwin Schrödinger of Zurich, Switzerland. Schrödinger's theory appeared to be quite different; it was based on differential equations (basic equations of calculus) rather than matrices, yet it solved the same problems. Physicists were much more familiar with differential equations than matrices, and it quickly became the more popular of the two theories. Within a short time, however, Schrödinger showed that the two theories were just different formulations of the same theory. They were equivalent.[5]

Einstein was repelled by Heisenberg's theory when he first encountered it. He was sure it was wrong, but his reaction to Schrödinger's theory was relatively positive. Still, when he studied the theory in detail, he was disturbed by its indeterminacy. To him it seemed incomplete.

A few years after Heisenberg put forward his matrix theory, he made another important discovery, now referred to as Heisenberg's uncertainty principle. He found that certain pairs of variables, such as position and velocity (or momentum), could not be simultaneously measured to a high degree of accuracy. If, for example, you narrowed in on the position of a particle and determined it to a high degree of accuracy, you could not simultaneously determine its velocity to the same accuracy. It is almost as if you had a microscope with two different focuses for position and velocity. If you focus on one, the other becomes fuzzy. Another pair of this type is energy and time.

There was also another difficulty with quantum mechanics, and it was related to the wave-particle dualism of the theory. Some experiments showed that the electron acted as a particle (in, for example, the photoelectric effect), and others showed that it acted as a wave. Bohr addressed this problem with his *principle of complimentarity*.[6] He pointed out that the wave-particle dualism was complimentary; the two aspects were exclusive, but they complimented one another. This particle-wave dualism is best brought out in the double-slit experiment.

THE DOUBLE-SLIT EXPERIMENT

The double-slit experiment was first performed by Thomas Young in the early 1800s. At that time it was an optical experiment, but we will see that it also has important implications for quantum mechanics. Let's begin by assuming that light is a wave and that we direct the light from a monochromatic source through a slit and allow it to strike a screen behind the slit. As expected, we get an image of the slit on the screen. Now replace the single slit with a double slit, where the slits are relatively close together. Young found that rather than two images, one of each slit, he got a series of bright and dark lines. He soon realized that this was caused because the two waves were interfering with one another. At some points they interfered constructively, giving rise to an enhancement of the light, and at others they interfered destructively, giving rise to light cancellation, or darkness.

But light is also composed of photons; in other words, it's made up of particles. So let's think of this experiment in terms of photons. Again, we'll begin with the single slit, but this time we'll do things a little differently. We'll release one photon at a time. Doing this, we see that we eventually get the same pattern with the single slit—an image of the slit. So far everything is okay. Now, let's go to the double slit, again releasing one photon at a time. Within a short time we see that we're getting the same image as before, namely, a pattern of bright and dark lines. If you think about it, this is strange. After all, a given photon goes up to the slit and passes through it. Now, however, it goes to a different position on the screen. But as it passes through this slit, how does it know that there is another slit a short distance away? It would obviously have to approach the slit, then check the area around it to see if there was a second slit nearby. If there was, it would go to one position on the screen, and if there wasn't, it would go to a different position.

So what's going on? We could, of course, place a small detector near the slits. Then, when a photon approaches, we could check to see which slit it went through. But according to a basic principle of quantum mechanics, if we disturb the system by measuring it, it goes to a different state, so we defeat our purpose.

The explanation of the dilemma is that light is both a wave and a particle, and its wave property has the ability to check the area around the single slit. It can spread out as it approaches the slits.

EINSTEIN AND BOHR

After quantum mechanics was discovered, Niels Bohr and his colleagues in Copenhagen, Denmark, struggled to understand the implications of the new theory. It was now clear that quantum mechanics was a radical departure from the usual classical theory, and some strange ideas had to be introduced to explain it. Their interpretation of the theory eventually became known as the Copenhagen interpretation. One of its central features was that a particle does not become real until it is measured. In other words, it doesn't exist until somebody detects it. This is generally in conflict with what most people believed, and still believe, so, as you might expect, Einstein had reservations about the idea.

Bohr was anxious to discuss his ideas about quantum mechanics with Einstein, and his first chance came at the 1927 Solvay conference in Brussels. All of the major physicists of Europe were at the meeting. Einstein did not present a paper, but he had many in-depth discussions with Bohr. For the most part, they consisted of Einstein presenting what appeared to be a shortcoming of the theory, and Bohr, after considerable thought, showing him how the controversy could be explained. Einstein and Bohr were close friends and had the highest regard for one another, so it was a friendly debate. It wasn't part of the proceedings of the conference, but it ended up overshadowing the formal proceedings. Indeed, it was later billed as the "battle of the Titans." Most of the discussions took place around the dinner table and in the evenings after the regular meetings were finished. In the end it seemed Bohr was able to counter all of Einstein's criticisms, but Einstein never conceded defeat and was not convinced.

In the years after the meeting Einstein spent a lot of time thinking about the theory, and when the next Solvay conference took place in 1930, he came prepared. He presented Bohr with a new paradox, one that surprised him so much he was unable to respond. The paradox was as follows.[7] Suppose you have a box that is filled with radiation, and on the box there is a small door through which radiation can be released. Furthermore, assume that the door is attached to a clock. Radiation has mass, so we can measure the box before and after the radiation is released, and as a result we know exactly how much energy has been released. At the same time, because there is a clock attached to the door, we know the exact time at which the energy (radiation) is released. But according to Heisenberg's uncertainty principle, we can't measure both energy and time to a high degree of accuracy. According to Einstein's thought experiment, however, we could.

Bohr was dumbfounded. He thought about the problem all evening long but couldn't come up with an answer. Finally, he decided to sleep on it. Maybe something would come to him when his mind was clearer. And sure enough, the next morning he had the solution, and as it turned out, it was an embarrassment to Einstein. He used Einstein's general theory of relativity to solve it. He pointed out that the box would have to be weighed, and there would be a slight change in its elevation during the weighing process. Its position in the gravitational field would change, and there was therefore a small uncertainty in the rate at which the clock ran. This uncertainty was all that was needed to bring back the uncertainty between energy and time.

Einstein knew when he was defeated, and he didn't contest the explanation. Still, he wasn't convinced.

EPR

Nazism was on the rise in Germany in the mid-1930s, and Einstein, as one of the best-known Jews in Europe, was a target. It soon became obvious that his life was in danger. He had been to the United States several times and was planning another trip in the fall, and this time he decided to stay. A job had been offered to him at the newly forming Institute for Advanced Study in Princeton, New Jersey, and he accepted it. After the turmoil in Europe, life was much more peaceful in Princeton, and he had time to look again at the difficulties in the foundations of quantum mechanics. He soon had a twenty-four-year-old assistant named Nathan Rosen, and together with him and a colleague, Boris Podolsky, Einstein looked again at the problem, and within a short time they had obtained an important result.

On May 15, 1935, Einstein, Podolsky, and Rosen published their results in a paper in *Physical Review*; it was titled "Can Quantum Mechanical Description of Physical Reality Be Considered Complete?" It is now referred to as the EPR paper, after the initials of the three authors.[8]

Three of the main issues in the paper were *reality*, completeness of a system, and locality. They began by giving a condition for reality, which was: "If without in any way disturbing the system, we can predict with certainty . . . the value of a physical quantity, then there exists an element of physical reality corresponding to the physical quantity." They then defined a *complete system* as one in which "each element of physical reality has a correspondence in physical theory." In other words, the theory is incomplete if it doesn't provide a picture of the world separate from our observations of it. Finally, they defined *locality* as, "the outcome of the measurement of certain quantum mechanical observations on one system are not immediately influenced by measurements made on a second system which is separated by some distance from the first." In essence, this means that the actions of a faraway observer, by themselves, can have no effect on something that happens locally. All of these statements seemed reasonable, and they were sure most scientists would not object to them.

The paper centered on a thought experiment that goes as follows. Assume two particles have just collided. Call them X and Y, and assume that after they collide, they fly apart and within a short time are thousands of miles apart. If we now measure the velocity (or momentum) of X, we immediately know the velocity (or momentum) of Y by conservation principles. In short, we have determined a property of particle Y without measuring it directly, but according to the Copenhagen interpretation, this is impossible. Particle Y doesn't exist until it is measured, according to this principle.

How do we explain this? According to Einstein, Podolsky, and Rosen, there are only two ways. Either quantum mechanics is incomplete, or there is some sort of "instantaneous communication" between the two particles. But if there is an instantaneous communication, it violates locality, since it would have to travel at many times the speed of light, which, according to relativity, is impossible. "Take your choice," Einstein said, and he was sure most scientists would agree that the theory was incomplete. This would mean there were "hidden variables" within the theory. In other words, there is a subtheory, which is presently beyond the sensitivity of our measuring techniques.[9]

As expected, the paper caused an uproar. Heisenberg and Pauli were outraged. "Einstein is up to his old tricks again," said Heisenberg.[10] They considered replying to it, but decided to leave it up to Bohr. Bohr was equally upset by the paper and decided immediately to reply to it. He was sure it would only take him a few days to come up with an appropriate response. But weeks passed and he still wasn't satisfied with his rebuttal. Finally, after six weeks he was ready; he published a short version of his challenge in *Nature* and a longer version in *Physical Review.* Bohr criticized Einstein's definition of reality and his statement, "without any disturbance to the system." To him they were unreasonable and erroneous; he argued that you could not speak of physical reality without the measuring process.

Most scientists agreed with Bohr, but Einstein had a strong ally. Schrödinger read the paper with delight, and he wrote Einstein a letter congratulating him. "You have publically caught the dogmatic quantum mechanics napping," he wrote.[11] He became so enthusiastic about the paper that he wrote a paper of his own.

ENTANGLEMENT

Einstein's argument centered on what is now referred to as entanglement. The two particles in the system he and his colleagues described were *entangled*. In a sense they were like a pair of trick dice. Let's assume we have several pairs of such dice. We'll assume that if we throw one of the dice of a pair and it comes up three, the second die will also always come up three. In another pair, if the first die comes up five, the second one also comes up five, and so on. In short, whatever the first die gives, the second will always match, and it doesn't matter how far the two dice are away from one another. I'm not sure there are trick dice of this type, but it helps illustrate our argument.

Now back to our discussion of entanglement. Before we get into the details, however, let's look at a slightly different approach to the problem. Einstein, Podolsky, and Rosen dealt with the position of the particle along with its velocity (or momentum). David Bohm of the University of London showed that we can argue the same thing in terms of what is called spin. George Uhlenbeck and Samuel Goudsmit showed in 1925 that most particles, including the electron, have a spin. From a simple point of view, we can think of it as similar to the spin of a spinning top.

But in this case there are only two spin states; a given particle can only be in one of two states: "spins up" or "spins down."

Now consider a system of two particles; we'll assume it is in its lowest energy state and has overall spin of zero. This can happen only if one of the two particles has spins up and the other has spins down. The two particles can flip back and forth, but when one flips, the other also has to flip, so that the overall spin is zero. These two particles are said to be entangled; in other words, if something affects one, the other is immediately affected. Now, let's apply the EPR argument to this system. Assume we create an entangled system with spin of zero and the two particles separate, with one moving thousands of miles away. If we measure the spin of the nearby particle, we immediately know the spin of the other because they always have to be opposite. But, again, we are in conflict with the Copenhagen interpretation.

It's important to remember that the particles can flip spin whenever they want. But if one flips, the other must also flip. So if we measure the nearby particle and find it is "spins up," the other particle must be "spins down." This seems to imply that there is some sort of instantaneous communication between the two particles. Einstein referred to it as "spooky action-at-a-distance."[12]

Is it possible? The only alternative is that locality is violated and that there are hidden variables in the theory. Einstein supported the idea of hidden variables, but he never tried to formulate a hidden variable theory. He believed that quantum mechanics would somehow be contained in an extension of his general theory of relativity, and all the problems would disappear when he formulated such a theory.

Einstein continued to work on the difficulties of quantum mechanics and published papers in 1948 and 1949, but they didn't have the impact of the EPR paper. Most people accepted Bohr's argument and assumed the difficulties had been overcome. Then in 1949 John von Neumann of the Institute for Advanced Study in Princeton presented a proof that hidden variables couldn't exist. It was contained in his book *The Mathematical Foundations of Quantum Mechanics*.[13] In short, he considered the possibility that quantum mechanics could be made into a deterministic theory where probability and chance played no role. He did this by checking to see if the theory contained additional variables that "are inaccessible to measurement and are therefore not subject to the restrictions of the uncertainty principle." He showed that it didn't.

BELL'S INEQUALITY

For several years there was little interest in the idea of hidden variables. Everyone assumed that von Neumann had laid the problem to rest. Then came John Bell.[14] Born in Belfast, Ireland, in 1928, Bell obtained his bachelor's degree from Queens University in Belfast in 1948 and his doctorate from the University of Birmingham in 1956. He specialized in quantum field theory and nuclear physics, but he soon became interested in the foundations of quantum mechanics. After reading von Neumann's proof, he showed that the assumptions von Neumann used were inappropriate and that his proof was invalid. His paper was titled "On the Problem of Hidden Variables in Quantum Mechanics." Einstein was not necessarily wrong, after all, and the points he had made in the EPR paper were more subtle than people realized. In a second paper Bell presented an inequality that could settle the problem once and for all. In particular, this inequality would allow a test to be set up, and it would settle the question as to whether quantum mechanics was incomplete. In particular, Bell showed that any "local" hidden variable theory cannot reproduce all the statistical predictions of quantum mechanics and that local hidden variables would produce results that contradicted the predictions of quantum mechanics.

Bell's paper presented an alternative: either local hidden variable theory was correct, or quantum mechanics was correct, but not both. And if quantum mechanics was the correct description of the microworld, then nonlocality is an important feature of the world, and instantaneous communication can take place.

What was crucial about Bell's inequality was that EPR was no longer a thought experiment. A real experiment could now be performed in the lab, but someone would have to work out the details of the experiment and design the needed apparatus.

SHIMONY, HORNE, AND CLAUSER

How could the experiment be set up? It would obviously take someone who was versed in both theory and experimental techniques, since the design of the experiment was a considerable challenge, and to do it you would have to understand the theory well. One of the first to become interested in performing the experiment was Abner Shimony.[15] Born in Columbus, Ohio, Shimony studied physics and math at Yale, obtaining a bachelor's degree in 1948. He then switched to philosophy and obtained a doctorate in philosophy in 1953. But his interest in physics continued, and in 1962 he also obtained a PhD in Physics from Princeton University. He taught at MIT for a while, then went to Boston University.

Shimony read both the EPR paper and Bell's papers and soon became fascinated with the foundations of quantum mechanics. He was sure Einstein had made an important point with his EPR paper and that Bell was right. An experiment to settle the matter was now critical. In 1968 he was approached by graduate student Michael Horne, who was looking for a thesis project. Shimony gave Bell's two papers to Horne to study and said, "See if you can expand on them and propose a real experiment to test what Bell is suggesting."

Horne went to work, and, together with Shimony, he spent most of 1968 and early 1969 working out the design of an experiment. They were almost ready to publish an article when they received a shock.

It's every scientist's nightmare to be scooped on a vital experiment just as he is going to press. And this is what happened to Shimony and Horne. They discovered that John Clauser of Columbia University was working on the same problem they were and had made some important advances. Clauser was a graduate student at Columbia, working on his PhD. He had read Bell's papers and was impressed, and he realized that a laboratory experiment could be set up, so he went to work designing one. He decided to model it on a famous Nobel Prize–winning experiment that had been performed many years earlier by Chien-Shiung Wu and Irving Shaknov. In this experiment an electron and a positron collided and annihilated one another with the release of two high-energy photons. Clauser talked to several scientists about his design, but no one seemed interested. They were all sure the experiment wasn't worth trying because Bohr had already won the debate. Finally, in 1969, however, Clauser got the breakthrough he was hoping for, and he was able to complete his design. He planned on presenting it at the American Physical Society (APS) meeting in Washington, DC, in the spring of 1969 and wrote an abstract for the *Bulletin* of the meeting.

Shimony and Horne had also planned on presenting their results at the 1969 APS meeting but missed the deadline for submission. They had also modeled their design on Wu and Shaknov's experiment. What was critical about this experiment was that the two photons emitted by the electron-positron annihilation had to be entangled. But these photons were very high energy, and it would be difficult to measure their spin. Lower energy photons would be needed. So Shimony and Horne devised an experiment using visible light photons. What they would have to do is measure the spin direction of the two entangled photons along an appropriate axis. They got their presentation ready for the APS meeting but missed the deadline. It didn't worry them, however; they were sure no one would be working on the same problem.

Shimony could hardly believe his eyes, however, when he read the *Bulletin*. In it was Clauser's abstract describing an experimental design similar to theirs. He called Horne and told him the news. They weren't sure what to do. Could they ignore Clauser's abstract and quickly publish their results? It didn't seem like a fair thing to do. After discussing it with several other scientists, they decided to contact Clauser directly, but they were fearful of his reaction. Many scientists are very protective of their turf, and they worried that he might be upset and think they were trying to scoop him. To their surprise, however, Clauser was pleased to hear from them, and after some discussion, they decided to collaborate. Furthermore, Clauser was pleased to learn that Shimony and Horne already had an experimentalist at Harvard University, Richard Holt, lined up to conduct the actual experiment.

The four men soon collaborated on a paper that was published in *Physical Review Letters*. It suggested an improvement in the design of the experiment, but the actual experiment had still not been performed. Clauser finished his PhD at Columbia and moved to the University of California at Berkeley, where he continued working on the design. With graduate student Stuart Freedman, he finally set up the experiment and was ready to go. Meanwhile, at Harvard University, Holt had also set up an experiment.

After two hundred hours of experimentation, Clauser and Freedman finally got a result. Einstein was wrong. Their results supported quantum mechanics and its strange action-at-a-distance communication. Einstein's idea of local realism appeared to be incorrect, and there were no hidden variables. But there was experimental error, and as a result, there was also some uncertainty. And strangely, at Harvard, Holt and his colleagues came to a different conclusion. Their results suggested that Einstein and his local realism and hidden variables was right. But again there was some experimental error. Indeed, after Holt heard about Clauser's result, he hesitated about publishing his own.

A more precise experiment was obviously needed, and it soon came.

ALAIN ASPECT AND OTHERS

Alain Aspect was born in France, not far from Bordeaux, in 1947.[16] He obtained his doctorate from the University of Paris in 1971 and immediately went to the Cameroon in Africa for three years. While in Africa, Aspect became interested in the foundations of quantum mechanics. He read both John Bell's papers and Einstein's EPR paper, and upon his return to the University of Paris in 1974, he began an intense study of the area. He read about Clauser's experiment in California, and about Shimony, Horne, and Holt's work, and decided he could design a better experiment.

Aspect decided to use a series of three experiments. Each was intricate in design and based on a different principle, but each experiment involved pairs of entangled photons. Over a period of many months he performed the experiments, and when the verdict was finally in, Einstein was shown to be wrong. Locality and hidden variables were defeated. Quantum mechanics was victorious.

The case was clinched by several other experiments that were performed about the same time. At Rochester University in the United States, Leonard Mandel used lasers in his experiment and again showed that quantum mechanics was correct. Yanhua Shih of the University of Maryland performed an experiment with the same result, and, finally, Nicholas Gisin of the University of Geneva got into the act. Using fiber optical cables that were approximately seven miles long, he showed that quantum mechanics was correct. There was now no doubt.

TELEPORTATION

One of the most exciting possibilities to come out of the phenomenon of entanglement in recent years is teleportation.[17] Everyone who has ever watched *Star Trek* knows all about teleportation. In almost every episode, Captain Kirk is on a planet somewhere beneath the *Enterprise* when he says into his mike, "Beam me up, Scotty." A few seconds later Captain Kirk mysteriously appears in a chamber aboard the spaceship. Until recently, no one in the scientific community took this seriously. It was the stuff of science fiction and probably wasn't possible in the real world. But in 1997 two teams of scientists showed that the quantum state of a particle could be transported. A single particle is, of course, a long way from a human being, but it's a first step.

Basically what happens in teleportation is that the state of a particle, in other words, its quantum properties, are scanned so that we know all there is to know about the particle. This information is then transmitted to another point, which can be a large distance away, and the information is transferred to another particle. We therefore end up with a particle that is identical to the original one. Anton Zeilinger and his colleagues of Vienna and Francesco de Martini of Rome both showed that this was possible. They were following up on an article by Charles Bennett in which he showed how such a teleportation was possible.

This was quite miraculous in that in the 1980s William Wootters and W. Zurek showed quite conclusively that a particle could never be cloned.[18] In other words, you couldn't make a particle that is identical to another one without destroying it. So you couldn't make numerous copies. In addition, this is in conflict with Heisenberg's uncertainty principle. According to it, you cannot measure position and velocity (momentum) at the same time to a high degree of accuracy. This means that you wouldn't be able to scan a group of atoms and extract all the information you needed for teleportation. To do it you would have to somehow bypass the uncertainty principle, and indeed, this is what Zeilinger and de Martini did.

One of the questions that comes up in relation to teleportation is: after you scan the atoms for all the information they contain, do you have to send the atoms themselves in addition to the information to the new site? This involves some deep issues and is generally not addressed in science fiction. We will assume for now that you don't have to send the atoms. If the new atom is identical to the old one, that's about all that is needed.

Still, how could we get around the problem of the uncertainty principle? Bennett and his colleagues suggested that we could use entanglement. If two particles are entangled, then separated, the particles always

Fig. 43: A simple representation of teleportation.

seem to know the state of their partners. This was a key part of their argument, but there was more. Let's assume we have two people; call them Pat and Mike, and each of them has one of a pair of entangled particles. We'll assume Pat's particle is in state X, and Mike's particle is in state Y, but neither Pat nor Mike knows the state of his particle. The only way they could learn it is by measuring it, and the measuring process would immediately change it.

We now have to be clever. Let's take another particle, call it Z, and allow Pat to measure it along with particle X; this creates an entangled state between particles X and Z. Because of this entanglement, Mike's particle responds immediately to it, and this gives him some of the information he needs. But it is not enough information for Mike to transform his particle Y over to state X. Pat must also communicate directly with him via wireless (or something of that sort), giving him the additional information he needs. When Pat does, however, Mike has a particle that is identical to X. In the process, of course, the X state at Pat's site is destroyed.

This is obviously a long way from teleportation of a human, but it shows us that it might be possible one day. There are obviously a lot of problems to overcome before that becomes possible. The amount of information needed to completely describe all the atoms and molecules in a human is mind-boggling. The body contains about 10^{28} atoms, and each would have to be described in complete detail. This would amount to more information than is contained in every book that has ever been written. And that's only one of the problems. It's obvious that it isn't going to be an easy task; nevertheless, it's exciting to consider.

FASTER THAN THE SPEED OF LIGHT

Entanglement seems to open a fascinating new possibility. If information is transmitted from one of an entangled pair to the other instantaneously, it is obviously going at a speed much greater than that of light. Is it possible that we could somehow feed information into one the particles of entangled pairs here on Earth and have it transmitted to distant points in the universe at greater than the speed of light? It turns out that this is impossible. As we saw in the last section, we can, indeed, transport information, but the information that is transmitted instantaneously is not sufficient; it must be supplemented with information that travels at the speed of light.

Scientists have, of course, speculated for years on particles that travel only at speeds greater than that of light.[19] They are called *tachyons*. But again there are difficulties. First of all, we haven't found any, and we're not sure they exist. If they do exist, they are in a world beyond ours, and at this point in time we don't know how to contact them.

EINSTEIN'S QUANTUM VISION

Again, it's obvious that Einstein started something that went well beyond what he expected. He didn't like the uncertainty associated with quantum mechanics, and many people thought he was just being stubborn. Indeed, after he argued day after day with Bohr about the foundations of quantum mechanics in 1927, his friend Paul Ehrenfest said to him, "Einstein . . . you are like your own critics of relativity."[20] But Einstein's criticisms were a result of deep thinking, and, as such, were extremely important. Furthermore, they made the proponents of quantum mechanics delve further into the meaning and implications of the theory. Einstein's EPR paper is one of the most important papers ever published in the area, and it spurred interest for years. It inspired so much interest that experiments were finally set up, and, as a result, Einstein's ideas were defeated. Nevertheless, science sometimes takes some strange twists, so it still might not be the end of the story.

Chapter 8

Superbombs

Of Einstein's many contributions to science, one of his greatest came back to haunt him in later life. It was the discovery of the formula that eventually ushered in the atomic age and led to the building of the atomic and hydrogen bombs. As a strong pacifist, Einstein hated war and everything it stood for, and he hated to think that he may have contributed to one of the greatest weapons of war that had ever been developed. He was a pacifist throughout his life, but when Hitler came to power, he realized that nothing short of military action would get rid of him. During this time he referred to himself as a militaristic pacifist. After World War II, however, he reverted to his previous stance of pure pacifism.

$E = mc^2$

The equation that gave us the atomic and hydrogen bombs, and nuclear reactors, is, without a doubt, the most famous equation ever written.[1] It is even familiar to people who know almost nothing about mathematics. It is a simple equation, yet it changed the world. It states that the energy (E) in a given amount of mass (m) is equal to the mass multiplied by the speed of light (c) squared (the square of a number means multiplying the number by itself). With the speed of light being 186,000 miles per second, its square is almost 35,000,000,000. Multiplying the mass by this obviously gives an extremely large number. And since Einstein's equation says that this energy is associated with any type of mass, a pound of anything is obviously going to produce a lot of energy. Fortunately, we can't convert mass directly to energy; it occurs only in certain nuclear reactions, and in these reactions, only a tiny fraction of the mass is converted to energy. Nevertheless, as we will see, this tiny fraction is enough to make a huge explosion.

Einstein published his formula relating energy and mass shortly after he published his paper on special relativity in 1905. It was, in a sense, an extension of it. Titled "Does the Inertia of a Body Depend on Its Energy Content?" it was only three pages long, but it was three pages that changed the world.[2] Initially, Einstein showed that if a body emits energy in the form of, say, radiation, then its mass decreases according to the above formula. A year later, however, he was able to show the reverse, and therefore the complete equivalence between mass and energy.

The discovery immediately explained a phenomenon that had been known for several years. Radioactivity had been discovered in 1891 by Antoine Henri Bequerrel of Paris. It is a phenomenon in which trillions of particles are given off by certain materials without an appreciable loss of mass.[3] Scientists wondered where all the energy was coming from. Einstein's equation explained it. Aside from this, though, there was little interest in Einstein's equation at first; it was just too abstract for most people.

MEITNER, HAHN, AND NUCLEAR FISSION

Although there were few advances over the next few years, the possibility that there might be large amounts of energy locked up in the atom was discussed. Many people wondered if it might one day be possible to unleash this energy. And although Einstein commented on it, he didn't take the idea seriously. He was sure that it wouldn't happen in his lifetime. "This idea has no direct support from the facts known to us so far. It is difficult to make prophesies, but it is within the reach of the possible.

For the time being, however, these processes can only be observed with the most delicate equipment," said Einstein.[4]

According to Einstein's equation, a single atom could produce only a small amount of energy, and therefore a large number of atoms would be needed to produce a large amount of energy. In effect, one atom somehow had to trigger others to release their energy. Such a process is referred to as a *chain reaction.* This chain reaction was, indeed, discovered, but not until several other important discoveries were made. The first of these discoveries came in 1934 when Enrico Fermi of Italy showed that projecting slow neutrons into a nucleus created new and heavier nuclei. He projected slow neutrons at all the heavy elements, including uranium, but he didn't take the time to analyze the chemical products fully. Otto Hahn, who worked at the Kaiser Wilhelm Institute in Berlin, heard of his experiment with uranium and decided to do it again and analyze the products carefully. He was assisted by Fritz Strassman, a graduate student who was working on his doctorate, and Lise Meitner, who had worked off and on with him for about twenty years.

Meitner was of Jewish descent, but she did not practice the faith and rarely thought of herself as a Jew. When Hitler issued a proclamation against the Jews, however, she knew she would have to leave Germany, but she would have to be careful because another proclamation had been issued that no academics could leave the country. To further complicate things, she didn't have a valid visa and knew she would arouse suspicion if she applied for one. She therefore got in touch with Niels Bohr in Copenhagen, asking if he could help. He arranged a position for her in Sweden, but he told her she would be on her own in getting out of the country. With an expired visa, she headed by train for the Dutch border; as the train approached it, she was stopped and questioned by German guards, but, eventually, to her relief, they let her pass. She later described the ordeal as the most terrifying of her life.

Once in Sweden, she became depressed. She was disappointed that she had been forced to leave at such a critical time. Hahn and Strassman were just beginning a series of experiments that might lead to an important breakthrough, and she was sitting far away in Sweden. She was pleased that Hahn had kept in touch with her. But he had a good reason: he knew that she was a much better theorist and had much better insight into experimental results than he did. He knew he might need her. And indeed, when he performed the first experiment just before Christmas in 1939, he got the shock of his life. Upon inspecting the products of the reaction, he found barium, and he knew that the barium nucleus was only about half as massive as the uranium nucleus. It seemed impossible that it could have been created in the reaction. Hahn was thoroughly confused and knew he needed Meitner's insight. He wrote to her in Stockholm.

Meitner was as surprised as Hahn by the development, and she had no explanation. It was close to Christmas, and she had been in touch with her nephew Otto Frisch, a physicist who was working at Bohr's institute in Copenhagen. She had asked him if he would like to spend Christmas with her. He had agreed, and they had decided to meet in the Swedish town of Kungälv. When Frisch arrived, Meitner showed him the letter from Hahn and asked his opinion. Frisch was as surprised as she had been and suggested that it had to be a mistake. Barium could not be produced in a reaction with uranium. But Meitner was certain that Hahn had not made a mistake.

That evening they went for a walk in the snow as they discussed the result. A model of the nucleus had recently been put forward suggesting that it was like a "liquid drop." Meitner wondered about the model. Was it possible that when the uranium nucleus absorbed the neutron, it began to oscillate? If so, it might eventually take on the shape of a dumbbell, and if this happened, the two halves would repel one another, and might split. Could this repulsion cause a splitting of the nucleus? Meitner asked herself. It would explain the presence of barium. It seemed like a long shot, but it was possible.

Stopping by a fallen tree, Meitner calculated the energy associated with such a splitting. She got two hundred million electron volts (an electron volt is the energy an electron gains in passing through a voltage difference of one volt). It was not a large amount of energy, but if many nuclei split, the cumulative effect could be large.

Meitner then considered the masses of the two barium nuclei and compared them to the uranium nucleus. There was a difference. Then, using Einstein's formula, she determined the energy equivalent of the difference in mass, arriving again at two hundred million electron volts. It was too much of a coincidence. The uranium nucleus had to have split. It was the only explanation. (Frisch later named the process *fission*, after a similar process in biology.) When Frisch got back to Copenhagen, he

explained the result to Bohr. Bohr was dumbfounded. He realized immediately that Frisch was right and encouraged him and Meitner to publish
the result as soon as possible. He was leaving for the United States the
next day and promised to keep it secret until they could publish it.

It wasn't a secret for long, however. Bohr's assistant didn't know of
Bohr's promise and told several people about it shortly after they got to
the United States. The news spread fast. Fermi, who was at Columbia
University, heard about it and realized he had missed a tremendous
opportunity several years earlier when he had bombarded uranium with
neutrons. He soon showed that in the process of fissioning, a uranium
nucleus would emit two neutrons, which would go on to fission two other
nuclei. A chain reaction would occur.

Several years earlier, Leo Szilard, a Jew who had fled to America
from Germany, had also conceived of a chain reaction. He had, in fact,
even taken out a patent on the process, but until now he hadn't been sure
the process would work. Now he realized that it would, and the idea
scared him. His biggest scare was that the Germans had the same information, and they would use it to build a bomb.

THE LETTER

Szilard was beside himself with anxiety. He had to stop the Germans. The
largest known source of uranium in the world was in the Belgian Congo,
and he was sure the Germans were aware of it. Earlier he had worked with
Einstein, and he remembered that he was a good friend of the queen of
Belgium. He decided to visit Einstein and encourage him to write to the
queen warning her of the German danger.[5] He phoned the Institute of
Advanced study at Princeton and was told that Einstein was vacationing
on Long Island. Along with Eugene Wigner of Princeton University, he
headed for Long Island. After some difficulty they finally managed to
locate his house. Szilard could hardly contain himself as he told Einstein
about the breakthroughs in fission and the possibility of a chain reaction.
Einstein had not kept up with nuclear physics, but he understood the
implications and realized that it was possible that the Germans could
build a bomb. He was soon as determined as Szilard to stop them, but he
didn't want to bother the queen of Belgium. He promised instead to write

a letter to a friend in the Belgian cabinet. They also talked about sending a letter to the White House.

Once he was back in New York, Szilard began thinking more and more about a letter to President Roosevelt. He was sure it would have considerable impact, but he had to find someone who had access to the president. Then he remembered Alexander Sachs, a vice president of the Lehman Corporation, who was one of the president's advisers. He quickly got in touch with Sachs, and to his delight, Sachs agreed to deliver a letter to Roosevelt.

Szilard composed two letters (a short one and a long one) to Roosevelt and took them to Einstein to sign. Einstein made a few changes and signed them, but fate intervened, and it was many weeks before Sachs could get to the president. Germany invaded Poland and World War II began; things happened so fast that Roosevelt had little time for visitors. Finally, however, Sachs got to see him and was relieved and pleased when Roosevelt took the threat seriously. He set up a committee to look into it. To Szilard's disappointment, however, the committee allotted only $6,000 for research.

HEISENBERG'S BOMB

Meanwhile, in Germany there was also considerable talk about a bomb. Werner Heisenberg had discovered quantum mechanics a few years earlier, and although he was still only in his thirties, he was now one of the most important scientists in the world. He knew the implications of Hahn and Strassman's discovery and was sure a bomb could be built. But there was almost as much reluctance from the German Weapons Bureau as there was in America. German officials were skeptical that such a bomb could be built; nevertheless, they gave Heisenberg the go-ahead to explore the possibilities. By February 1940 he had a report ready that gave the outline of how such a bomb could be constructed. The first step would be to build a slowed-down version of the bomb—a nuclear reactor.

One of the major problems in building a reactor was slowing the neutrons. There were two known methods for doing this. Graphite was an excellent moderator, as was heavy water, or deuterium. Heisenberg decided to go with heavy water. He ordered as much uranium as he could get and formed it into a "pile" with heavy water dispersed in containers throughout it. Several measuring devices were also set up around the pile. According to Heisenberg's calculations, a chain reaction should occur with the amount of uranium they had. But when the test was made, nothing happened. Heisenberg went back to his calculations and discovered that more uranium would be needed. He ordered more.

THE AMERICAN BOMB

Things were going too slowly for Szilard. America was dragging its feet. Aside from a few committee meetings and a small allotment of money, almost nothing had been done. He went to Einstein again. They had to write another letter to Roosevelt. In this letter Einstein mentioned that he had recent information that the Germans had a secret project related to uranium, and he was sure they were planning to build a bomb.

Still, little was done. But Frisch and others were also working on the British government, and it was becoming alarmed by developments in Germany and sent a report to the United States. Finally, in the fall of 1941, things began to roll. Money was allotted for the building of a nuclear pile at the University of Chicago. Enrico Fermi was to head the program. This pile would be similar to the German one, but the moderator would be graphite rather than heavy water. Construction began in November 1942. The idea was to build the pile large enough so that the chain reaction would just barely go critical and could be controlled. The fuel was uranium-235, which had to be separated from natural uranium, a mixture of uranium-235 and uranium-238.[6]

On December 2, 1942, the pile was ready. Everyone assembled on a balcony overlooking it. One of the moderating rods was set so that when it was pulled out, the pile would go critical. At 3:25 that afternoon Fermi ordered the rod pulled out. The Geiger counters around the pile began clicking, then suddenly they began to roar. The pile had gone critical! A bomb could be built.

TO NORWAY

Meanwhile, the British were becoming increasingly nervous about the German bomb. Bohr met with Heisenberg at a joint scientific meeting between Denmark and Germany in October 1941. Bohr knew that the Germans would be watching carefully, but he was anxious to talk to Heisenberg about the bomb. He knew that Heisenberg was the head of the German project, and he had to know what they had achieved. Strangely, even though he knew he was breaching security, Heisenberg discussed the project with Bohr. He even made a rough drawing of the pile and gave it to Bohr.

Bohr was in shock. The Germans had to be stopped. He passed the information on to British intelligence, but they had already been monitoring the program from the beginning, and they also felt that something had to be done. The weak spot in the program was the German dependence on heavy water from Norway. Germany had occupied Norway and now had access to it. The heavy water factory was located in a mountainous region about ninety miles from Oslo. It was almost impenetrable, with the only access to the factory being a bridge across a deep ravine. Furthermore, the factory was guarded by several hundred troops.[7]

The British decided to attack it using two teams of highly trained commandos. There were thirty in all, and they planned to use two gliders, towed by two bombers. The gliders were to land near the factory, and the troops were to somehow get inside it and sabotage it. But tragedy struck. The weather did not cooperate, and they were caught in a blinding snowstorm. The bombers lost their way in the storm and had to release the gliders early; both of them crashed, with the loss of several men. Most of the others were eventually captured.

The failed mission was a shock to the British. They were reluctant to plan a second mission but were still determined to do something. After considerable discussion they selected six Norwegians who were familiar with the region, hoping that they would be more successful. The British army, however, was skeptical; after all, the army had sent its best troops, and they had failed. The six Norwegians were parachuted into Norway where they met up with the Norwegian resistance. It took them several weeks to ski to the factory, and then they had to decide how to get inside it. They couldn't go across the bridge; it was too heavily guarded. This meant they had to climb down the hundred-foot ravine that it crossed, and

Fig. 44: The heavy water plant in Norway (about ninety miles from Oslo).

indeed this is what they did. And, by a stroke of luck, an engineer had told the Norwegian resistance about a cable duct on the side of the factory that could be used to get in.

In ten minutes the charges were set and the Norwegians were racing away from the site. From the outside the blasts were hardly heard, or seen, but they had been carefully placed, and they did what they were meant to do. The heavy water poured out of the building, and the apparatus for making it was destroyed. All the Norwegians escaped.

THE MANHATTAN PROJECT

With the success of the nuclear pile at the University of Chicago, the atomic bomb project shifted into high gear.[8] Brig. Gen. L. R. Groves was selected as the commanding officer of the project. Several problems had to be considered. First of all, a sufficient amount of uranium-235 had to be separated, and a plant was set up at Oak Ridge, Tennessee, to do this. Furthermore, a central laboratory had to be set up where research and engineering could be carried out, and it would have to be isolated so it could be kept secret. A director for this facility would also have to be appointed. The names of several people were suggested, but the one of most interest to Groves was J. Robert Oppenheimer of the University of California and Caltech.

Groves put forward Oppenheimer's name to the military policy committee, but they were not happy and suggested he come up with someone else. Oppenheimer had no experience directing people, and he was controversial. Groves looked into other possibilities, but he soon decided that Oppenheimer was the best man for the job. He resubmitted his name to the committee, and they reluctantly accepted it.

A site was selected near Jemez Springs, New Mexico. It was named Los Alamos, after one of the nearby canyons. Construction began in early 1943. The next problem was personnel, and Oppenheimer was soon scouring the country for the top scientists. Within months he had a first-rate team, including many Nobel Prize winners, and over the next two years they worked day and night to beat the Nazis to the atomic bomb. By early 1945 all the problems had been overcome, and on July 16 the bomb was tested at the Alamogordo Air Base, about fifty miles from the city of Alamogordo. The test was successful. The bomb was ready to drop.

THE CONTINUING GERMAN THREAT

In early 1944 British intelligence learned that the heavy water plant in Norway had been repaired and was producing even more heavy water than it had earlier. Furthermore, the plant was now under even greater guard, so there was little chance that anyone could get through to sabotage it again. Then they learned that a large shipment of heavy water was scheduled to be shipped to German in mid-February. Furthermore, it had to be taken on a ferry across a lake near the plant. The Norwegian underground waited for its chance. The night before the ferry was scheduled to leave, several Norwegians were able to plant explosives on it. The following morning the explosives went off, taking the ferry, along with its valuable cargo of heavy water, to the bottom of the lake.[9]

This was a disastrous blow to the German program, but it wasn't the only thing that caused delays. Hitler had never been fully committed to the bomb and had not given it a lot of support. Near the end of the war, because of extensive bombing, the project was moved to the town of Hechingen in southern Germany. The lab was set up in a cave where it could not be seen by overhead observation planes. But in the end the pile never did go critical, and work on the bomb itself did not begin. Despite their two-year lead at the beginning of the war, the Germans lost the race to the bomb.

EINSTEIN'S ROLE

Einstein was not directly involved in the making of the bomb. He never even visited Los Alamos, but he was quite aware of the project and was kept abreast of developments by several people. Niels Bohr, who was associated with the British side of the Manhattan Project, visited Einstein in December 1943 and no doubt brought him up to date on the bomb's progress. His old friend Otto Stern, who participated in the project, also visited him several times. It is perhaps ironic that the FBI considered Einstein too much of a risk to be associated with the project. FBI director J. Edgar Hoover even went as far as keeping a file on him for several years, but he never did find anything.

Einstein did, however, contribute to the war effort in several ways. In 1943 he was called on by the navy to work as a adviser on explosives.[10] Every two weeks or so he was visited by explosive experts; they brought him problems related to the explosion of torpedoes and so on. Einstein was happy to be of assistance and had solutions to their problems ready when they visited him. But there's no indication that a major breakthrough occurred in the area because of Einstein's work. Einstein did, however, make a large financial contribution to the war. An organization that auctioned off rare manuscripts, with the proceeds going to the war effort, asked Einstein for his 1905 manuscript on special relativity. He was unable to oblige them because he had destroyed it shortly after it was published, but he did offer to write it out again and donate the proceedings. At auction, it fetched six million dollars. Another manuscript that he was working on brought five million dollars.[11]

Einstein was vacationing on Saranac Lake in the Adirondacks when he heard the news that the United States had dropped an atomic bomb on Hiroshima. His only comment was "Alas, Oh my God!"[12]

THE HYDROGEN BOMB

Einstein was severely depressed when he heard a few years later that an even more powerful bomb was being built. It was also based on his energy-mass equation, but the nuclear process that it used was different. As we noted earlier, the atomic bomb used fission to achieve its energy. But there is another similar process that gives off even more energy, and it is the process that fuels the Sun. It is well known that the Sun and stars are generating unbelievable amounts of energy. The total energy emitted by the Sun is equivalent to a half million billion billion horsepower. This is equivalent to a billion atomic bombs going off every second. Earth, of course, receives only a tiny fraction of this energy.

There is no uranium in the Sun; it is composed mostly of hydrogen and helium, so its energy must be associated with these elements. And indeed the fuel of the Sun is hydrogen; energy is produced when hydrogen atoms come together, or fuse, to create helium. The process is therefore called *fusion*. Four hydrogen nuclei go into the making of helium, but if you weigh four hydrogen nuclei and compare their masses to that of the helium nucleus, you see that they are slightly different. In the process of coming together, a small amount of mass is converted to energy, an amount that can be calculated using Einstein's equation.

The exact process of conversion from hydrogen to helium was worked out by Hans Bethe of Cornell University, for which he later received the Nobel Prize. It is not a straightforward process, but involves several steps. From a simple point of view, two hydrogen nuclei (which are just protons) collide and join to form a deuterium nuclei, which is a heavy form of hydrogen. (It is referred to as an isotope.) Another hydrogen nuclei then strikes the deuterium nuclei and creates a nuclei of tritium, another isotope of hydrogen. In the final step, two nuclei of tritium collide to form helium.

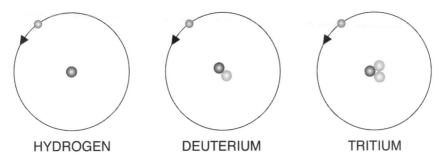

HYDROGEN   DEUTERIUM   TRITIUM

Fig. 45: The hydrogen, deuterium, and tritium atoms.

Knowing the details of the process that is taking place in the Sun, and knowing that it produces a large amount of energy, scientists immediately began to look at it as the basis of an even more powerful bomb than the atomic bomb. We now refer to it as the hydrogen bomb, but while it was being developed, it was frequently referred to as the "super."[13]

The sequence of reaction that takes place in the cycle that converts hydrogen to helium in the Sun takes almost five million years to complete and obviously wouldn't be helpful in building a bomb. It works in the Sun because there is so much hydrogen. What would be needed for a bomb would be a reaction that takes place in a tiny fraction of a second. Scientists were therefore faced with two problems. First, they had to give the hydrogen nuclei sufficient energy so that they could penetrate the repulsive electrical barrier around other nuclei. Second, the reaction had to go fast enough—preferably within a millionth of a second. The first of these was solved by the atomic bomb. For a brief instant after the bomb explodes, the temperature near its core reaches twenty to thirty million degrees. This is high enough to give the hydrogen nuclei enough energy to penetrate the electrical barrier of other nuclei.

The second problem, namely, the speed of the reaction, was solved by looking at hydrogen's two isotopes, deuterium and tritium. There are several reactions involving deuterium and tritium that take place very rapidly. For example, when a deuterium nuclei hits a tritium nuclei to form helium, the process occurs in .0000012 second, and it generates an enormous amount of energy. One of the problems with this is that we have to separate deuterium and tritium from ordinary hydrogen (which is a mixture of the three isotopes), and that isn't easy. Ordinary water is used as the source of the hydrogen. But only about one atom in five thousand in ordinary water is deuterium, and only about one in a billion is tritium. Another problem is that the hydrogen (deuterium and tritium) must be held in contact with the exploding atomic bomb for a time long enough to heat the hydrogen to twenty million degrees.

Several people were already thinking about the "super" in 1945 when plans were being made to test the first atomic bomb at Los Alamos. But it was several years before any action was taken. Interestingly, President Truman never heard that it was possible to build such a bomb until late 1949. Edward Teller spearheaded much of the early effort to get a hydrogen bomb project initiated.[14] He proposed a design for the bomb and tried to get funding for it. There was, however, considerable reluctance to building a bomb that was even bigger and more destructive than the atomic bomb. Harvard president James Conant was strongly against it, and Oppenheimer, who had headed the Manhatten Project, was also reluctant to get involved in it. Conant was sure that building such a super-bomb was morally wrong, and Oppenheimer soon shared his opinion.

According to many people, including Teller, however, there were strong reasons for building it. One of the strongest was that the Russians were also involved in thermonuclear research, and it was possible they would produce such a bomb within the next few years. According to Teller, "If the Russians continue to make actual progress faster and if we lose the atomic armament race . . . our situation would be hopeless."[15] Despite the opposition, most scientists were strongly in favor of developing the bomb as quickly as possible, and in late 1949 President Truman gave the go-ahead.

Numerous calculations were needed, but fortunately large supercomputers were now coming into their own. One of the first to be completed was MANIAC, and it was used extensively in performing the necessary calculations. As was the case with the atomic bomb, most of the work was done at Los Alamos.

From a simple point of view, the bomb works as follows. An atomic bomb is blasted off, which acts as a trigger. This blast produces temperatures of about twenty million degrees, which cause the mixture of deuterium and tritium to react, creating even higher temperatures. More and more of the hydrogen fuses, and a tremendous explosion follows. All of this occurs in a tiny fraction of a second.

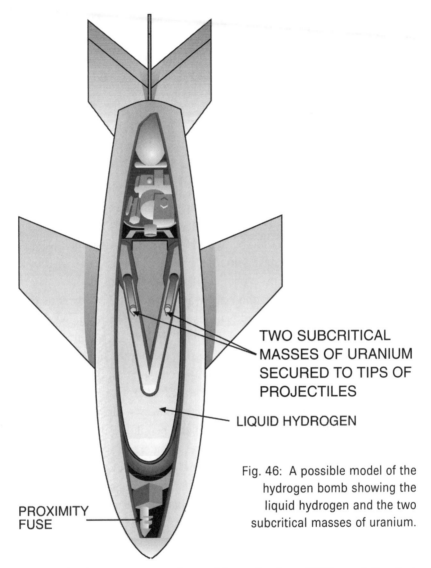

TWO SUBCRITICAL
MASSES OF URANIUM
SECURED TO TIPS OF
PROJECTILES

LIQUID HYDROGEN

Fig. 46: A possible model of the
hydrogen bomb showing the
liquid hydrogen and the two
subcritical masses of uranium.

PROXIMITY
FUSE

The scientists overcame the problems by late 1950, and the first hydrogen bomb was exploded in the spring of 1951 on the island of Eniwatok, about three thousand miles west of Hawaii in the Pacific Ocean. On March 1, 1954, a bomb that could be dropped from a plane was exploded on Bikini Island in the Pacific. It had a power of fifteen megatons (fifteen million tons of TNT). But the Soviet Union wasn't far behind. It tested its first hydrogen bomb on August 12, 1953. It was less powerful than the American bomb, but the Soviets soon had a more powerful one. And the race was on to build and stockpile more and more bombs. Unlike the atomic bomb, which has limited power, there is, in theory, no limit to the power of the hydrogen bomb. Add more hydrogen, and you get a bigger blast.

EINSTEIN'S REACTION

Einstein is sometimes referred to as the "superfather" of the atomic bomb (Oppenheimer was considered the "father"), because of his famous equation that was at the basis of it and because he alerted President Roosevelt to the possibility that it could be built. His formula is also the basis of the hydrogen bomb, but he played no role in the building of it. He had by now resumed his prewar role as a pacifist. The hydrogen bomb severely troubled him, and he was sure the only way to avoid its use was to form a world government. The United Nations was formed after the war, but this wasn't quite what Einstein had in mind. With two superpowers—the United States and Russia—both possessing the bomb, and each very wary of the other, there was little in the way of cooperation between them.

Einstein worked hard to push his idea of a world government by signing several documents and promoting his views. He was determined to completely abolish war, but to most of the people around him, his effort appeared to be a hopeless. His ideas were considered naive to many, and his efforts seemed to bring few results. At times he came close to despairing for humanity. "If all efforts are in vain and mankind ends in self-destruction, the universe will not shed a tear over it," lamented Einstein.[16]

Chapter 9

Other Einstein Insights

E instein's major contributions were so great that they completely overshadowed most of his minor contributions. Referring to them as "minor" contributions, however, hardly does them justice. Most of them would be crowning achievements for any other scientist. Three of these contributions were made after receiving a letter from an Indian physicist, Satyendra Bose, in 1924. They are: Bose-Einstein statistics, the prediction of superfluidity, and the prediction of the Bose-Einstein condensate. Among Einstein's other contributions was the discovery of the process that led to the invention of the maser and laser, and the prediction of the existence of photons, which eventually led to a large number of energy-saving devices. Even while working on his doctorate, he was making important discoveries. In his doctoral thesis he showed theoretically that atoms and molecules had to exist. And shortly after that he explained the strange phenomenon that had been discovered by Robert Brown, called Brownian motion.

Interestingly, Einstein was also responsible for a number of inventions. He couldn't compete with Thomas Edison in this respect, but he did enjoy tinkering. This went back to the years he spent in the patent office in Bern—years that he enjoyed. His main inventions were related to cooling; he took out patents for several designs of refrigerators, but he was also involved in several other devices.

BOSE

Born on January 1, 1894, in Calcutta, India, Satyendra Nath Bose was educated at University College in Calcutta, where he obtained his masters degree in 1915.[1] Bose taught and did research at University College until 1921 when he left for Dacca University. Like many scientists, he had an obsession for understanding things thoroughly. "As a teacher who had to make these things clear to his students . . . I wanted to know how to grapple with the difficulties in my own way," Bose wrote.[2]

Soon after moving to the University of Dacca, Bose became interested in Planck's derivation of his radiation law. In 1900 Max Planck of the University of Berlin had shown that the radiation curve for a heated object could be explained by introducing the idea that emission and absorption of heat took place in chunks that he called "quanta." Prior to this, it had been assumed to be a continuous process. Several people, including Einstein, had derived Planck's formula in a different way. But Einstein was not satisfied with his derivation; it contained both classical and quantum concepts, and this bothered him. He would have preferred a derivation that relied only on quantum concepts.

Bose read about Einstein's dissatisfaction and decided to look into the problem. Many years earlier Einstein had introduced radiative quanta (photons) and had suggested that the radiation given off by a heated object was photons of various frequencies. Bose decided to use this idea in his derivation. He also noticed that Peter Debye of Zurich had used statistical mechanics in deriving the formula, but like all the other derivations, it contained both classical and quantum concepts.

Bose decided to use statistical mechanics, but he was sure the standard statistical procedure was not appropriate for his derivation. The only known particle statistics at the time was Boltzmann statistics, in which each particle was labeled; in other words, it was different from all other particles, and it was placed on energy levels according to its individuality.

Fig. 47: Satyendra Nath Bose.

Bose made the assumption that photons were not distinguishable. They were all the same, and, because of this, they would be placed on the energy levels in a different way. It was a strange, new approach, but it seemed reasonable, and he was pleased when he found that he was able to derive Planck's formula using it. Of particular importance, the derivation employed no classical concepts.

Late in 1923 Bose submitted his paper to *Philosophical Magazine* in England, but to his disappointment it was rejected. He didn't let this deter him since he was sure the result was important. Furthermore, he knew that Einstein was interested in the problem, so he decided to send the paper to him. Writing to Einstein, he said, "I have ventured to send you the accompanying article for your perusal and opinion. I am anxious to know what you think of it. . . . I do not have sufficient German to translate the paper. If you think the paper is worth publishing, I shall be grateful if you would arrange for its publication in *Zeitschrift für Physik*."[3]

Eleven days later he sent Einstein a second paper that continued the ideas in the first paper. In July 1924 he got a postcard from Einstein acknowledging them.

EINSTEIN'S RESPONSE

Einstein had indeed been working for several years on the problem that Bose had solved. He had derived Planck's formula several ways but was not satisfied with any of them. He was therefore pleased when he received the paper, and he recognized its importance immediately. Furthermore, he saw several important implications of Bose's new statistical method. Bose had applied his statistics only to photons; Einstein saw that it also could be applied to atoms and molecules. Unlike Bose, who thought of it as a trick that helped his derivation work, Einstein saw it as a new form of statistics, distinct from Boltzmann statistics. We now refer to it as Bose-Einstein statistics.

Even before Bose's paper was in print, Einstein presented a paper titled "Quantum Theory of the Single-Atom Ideal Gas" to the Prussian Academy.[4] A few weeks later he presented a second paper to the academy. The most important result of these papers was showing that Bose's ideas could be extended to a quantum statistics that was valid for both radiation and matter. It was later shown that Bose-Einstein statistics did not apply to all types of particles, however, only those with a particular spin. Nevertheless, it was an important discovery.

Einstein also used Bose statistics to look at what would happen to gases at very low temperatures. He saw that their viscosity or "stickiness" would disappear. This phenomenon was soon called "superfluidity," and four years later, in 1928, it was demonstrated by Willem Keeson in Leyden. Indeed, Einstein went further than this. He looked at the properties of atoms at extremely low temperatures—all the way down to absolute zero. Absolute zero is the lowest possible temperature in the universe; particles have no motion at this temperature. He found a strange result: atoms would become "superatoms." In short, the quantum "wave packets" that represented the atoms would grow, and as they did, they would begin to overlap one another, until, finally, at absolute zero, the packets would coalesce into a single macroscopic wave packet, which we would see as a superatom. This was a new form of matter, one quite different from the known forms: solid, liquid, and gas.

At the time there was no way of cooling anything close to zero degrees absolute, so the discovery did not attract a lot of attention. It was called a Bose-Einstein condensate (BEC), and it was several years before scientists began to accept the fact that it could exist.

Bose was pleased to hear that his papers were published, and with the help of the postcard from Einstein, he was able to get permission from his university to travel to Europe for advanced study. He sailed in September 1924, but he did not go directly to Germany. He could not speak or write German, but he was conversant in French, so he stayed in Paris while he studied German. He soon made contact with Einstein in Berlin, however, and a few months later he visited him. Einstein introduced him to many of the scientists at the University of Berlin, and he suggested two problems for Bose to work on. But with all the activity and so on, Bose made no progress on them. He returned to India in the summer of 1926.

BOSE-EINSTEIN CONDENSATE

Einstein's prediction of a "condensate" was, without a doubt, one of the most interesting predictions to come out of the Bose-Einstein collaboration.[5] But with no way of cooling things to absolute zero, there was little interest in it. It was not until 1985, when Steven Chu of Stanford University, Claude Cohen-Tannoudje of College de France, and William Phillips of the National Bureau of Standards showed that laser light could be used to cool atoms, that interest was sparked. It seemed possible that, using this new technique, it might be possible to reach absolute zero. Chu, Cohen-Tannoudje, and Phillips received the Nobel Prize for their discovery in 1997.

In 1989 Eric Cornell and Carl Wieman of the Joint Institute for Laboratory Astrophysics at the University of Colorado decided to apply the laser technique to see if they could produce a BEC. Others had tried hydrogen and had problems. They decided to try rubidium gas. They soon saw that laser cooling by itself would not allow them to accomplish their goal, so they decided to use magnetic trapping and evaporative cooling in conjunction with it. In magnetic trapping, strong magnetic fields are used; each of the atoms has a tiny magnet associated with it, and the strong external field acts on these magnets to control the atoms' motion. In evaporative cooling, the more energetic (hotter) atoms are allowed to escape the container, leaving only the cooler ones.

Cornell and Wieman used a small glass container surrounded by several coils of wire in their experiment. They began by evacuating the container and introducing a small amount of rubidium gas. Six laser beams were then shone on the gas from different angles. The frequency of the laser light was adjusted so that the rubidium atoms would absorb and emit photons from the beam. The idea was to get the atoms to absorb as many photons as possible that were traveling in the direction opposite to the atom's motion. This absorption would apply a small "kick" to the atom, slowing its motion slightly, and with enough of these kicks, the atom could be cooled to very close to absolute zero. It was important, however, to apply these kicks in exactly the right way; they had to be opposite the motion of the atoms. Also, it was important to keep the rubidium atoms away from the walls of the container since they were at room temperature. This was done by applying a weak magnetic field; the wires around the cell applied this field.

Using this technique the atoms near the center of the container were cooled to a temperature of forty millionths of a degree above absolute zero, but, amazingly, this wasn't low enough to get a BEC. Magnetic trapping and evaporative condensation were therefore applied. As we saw earlier, for this, a strong magnetic field was needed. Cornell and Wiemer built their apparatus in 1995 and observed the first BEC late that same year. They took a snapshot of the BEC using laser light. It was seen as a

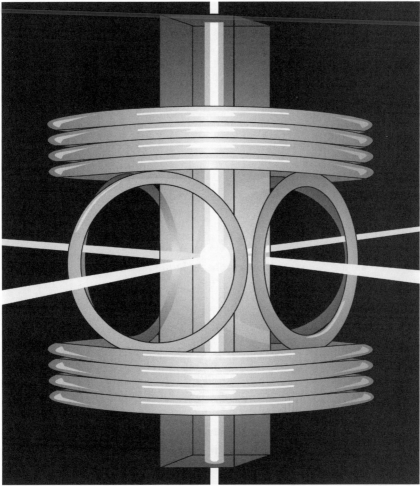

Fig. 48: Cornell and Wieman's apparatus for producing a "condensate."

small lump or droplet in the bottom of the container. The droplet was surrounded by normal gas, so it looked a little like the pit in a cherry.[6]

Several labs have now produced BECs. With it becoming increasingly common, we immediately wonder what it is good for. How will it
be helpful to us, or will it remain just a scientific curiosity? To answer
this, it is best to look at its analogy, the laser. In a laser beam all the photons are of the same frequency, and they are all in phase. Because of this
there is almost no scattering of photons out of the beam, and we have

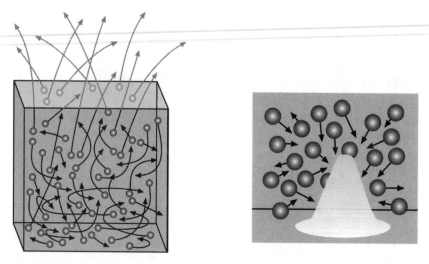

Fig. 49: The atoms before the condensate formed are shown on the left.
The forming condensate is shown on the right.

amazing control over it; we can, in fact, concentrate it so that it acts as a
welding, or cutting, torch. This is only one application of the laser. As is
well known, there are many other applications.

In a BEC all the atoms are in phase; in essence, it is like a single
atom, and so we have considerable control over it, just as we do a laser
beam. Its usefulness will likely therefore center on this control. So far,
however, the discovery is so new that it has not been developed techno-
logically.

EINSTEIN'S INVENTIONS

Einstein's "discoveries" are, of course, well known, but he was also
responsible for several interesting inventions. It is perhaps natural that he
had an interest in inventing. After all, he worked for nine years at the
patent office in Bern. Many people who worked at the patent office found
the work boring, but Einstein enjoyed it; he was continually amazed at
how basic physical principles could be used to make practical devices. He
was too busy during the first few years after he left the patent office to

Fig. 50: Leo Szilard.

dabble in inventions, but by 1920, when he was just over forty, his interest in the area was rekindled by one of his students, Leo Szilard.

Szilard was born on February 11, 1898, in Budapest, Hungary.[7] Upon graduation from university in Hungary, he came to the University of Berlin to work under Einstein. Einstein recognized his brilliance immedi-

ately and assigned him the dissertation topic of extending classical thermodynamics to fluctuating systems. Einstein was impressed with how he
handled the problem and invited him to his house several times. Although
they usually talked about physics, Szilard was interested in inventions,
and this appealed to Einstein.

One day Einstein was reading the newspaper when he noticed an
article about a family who had been killed by the toxic gases that leaked
from their refrigerator. He was disturbed and mentioned it to Szilard.
Together, they looked at the design of the refrigerator and decided that
gas could leak from several different places, and most were associated
with the moving parts of the refrigerator.[8] The main difficulty, they
decided, was the mechanical pump in the device. At that time (and now)
all refrigerators used mechanical compressors. A refrigerant gas was compressed by a pump; in the process it liquefied as its heat was discharged
to the surroundings. The liquid was then allowed to expand, and as it did,
it cooled, thereby cooling an interior chamber in the refrigerator. Einstein
and Szilard decided to design a refrigerator that had no moving parts. In
their first design, they used a natural gas flame to drive the cooling cycle.
But once they got going, ideas flowed, and over the next few months they
came up with three different designs, all based on a different physical
principle. None of the three had any moving parts.

With Einstein's experience in the patent office, they were able to get
patents for the devices relatively easily, and in late 1926 they sold their
first invention to a company in Berlin. Szilard immediately began building a prototype, but within months the company went bankrupt. A short
time later, however, they were able to sell it to another company.

One of their more unique designs was a cooler for a glass of beverage.
Again, it had no moving parts, using only the water pressure from a tap. It
was sold to a company in Hamburg, but problems soon developed. The prototype worked well, but when they attempted to market it, Germany's hap-
hazard water system created problems. The water pressure varied so much
from city to city that the device could not be adjusted properly.

The most successful invention of the Einstein-Szilard collaboration
was an electromagnetic refrigeration pump.[9] It was a fully functional pump
with no mechanical moving parts that relied on an electromagnetic field and
a liquid metal. It was bought by AEG of Germany (the German equivalent
of General Electric). Over a period of two years a prototype was built. It

worked well, and although it was noisy, it was quite efficient. It went into operation in July 1931. Unfortunately, the Great Depression had a serious effect on AEG, and it decided not to market the new model. It was, however, later used in nuclear reactors. Soon after this, Freon was discovered in the United States. It was not toxic, so the problem that Einstein and Szilard were trying to overcome was solved. It was later shown, however, that Freon damaged the ozone layer above Earth, so it became illegal.

While Einstein and Szilard were working on their inventions, the Nazis gained power, and because he was a Jew, Szilard decided to leave Germany, so the collaboration ended. Einstein later fled the country, since he was also Jewish. Over the seven years they worked together, however, Einstein and Szilard filed forty-five patent applications in six countries, which was no mean feat. In the following years Einstein was involved in several other inventions, including a hearing aid and a gyrocompass.

MASERS AND LASERS

In 1916 Einstein published three papers on quantum theory. The first contained an elegant new derivation of Planck's law of radiation, and it introduced the concept referred to as a *transition probability*. In 1913 Bohr had introduced his model of the atom, which included energy levels, and he had assumed that electrons jumped between these levels, but he did not go into the details on how the jumps occurred. Einstein showed that two types of jumps, or transitions, were possible. He referred to the first as *spontaneous emission* transitions. This type of transition occurred when an electron was in an upper (or excited) energy level and jumped to a lower one. In the process a photon was released. The second type of transition was referred to as *stimulated* or *induced emission*. In this case a photon strikes an electron that is in an excited state, but the electron does not absorb the photon, rather it is stimulated to fall to a lower level, and in the process it emits a photon. There are therefore two photons associated with the process.[10]

Another way of looking at this is that a photon strikes an atom that is in an excited state. The atom, in turn, emits a photon like the first one that travels in the same direction. There are, then, two photons traveling in the same direction. Each of them can hit other excited atoms, both of which

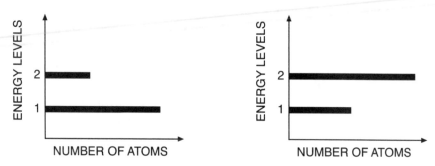

Fig. 51: A normal distribution of atoms is shown on the left.
A population inversion is shown on the right.

emit photons. And so on. The end result is that the number of photons is amplified. This is where the word "laser" comes from; it is short for "light amplification by stimulated emission of radiation."

There is a problem here, however. In order to excite the atom, we have to hit it with a photon. So in the end it takes two photons to get two photons. To overcome this, we need what is called a *population inversion*. Normally, an atom has most of its electrons in what is called the ground state, but if we could create a state in which most of the atoms were excited, we would have a population inversion. As it turns out, we can get this by pumping energy (e.g., shining light) into the atoms. In this case, when an excited atom gives off a photon, this photon hits other atoms, causing them to give off photons, until, finally, a cascade of photons is created.

Joseph Weber pointed out in 1951 that an amplification device could be built using a population inversion. A few years later the first maser (short for "microwave amplification by stimulated emission of radiation") was built by Charles Townes and several of his students at Columbia University.[11] The photons in this device were microwave photons. With the development of the maser, scientists began to think of a similar device in the optical region of the spectrum. Townes, of course, was one of the first to work on the project. Shortly after he began, he teamed up with Arthur Schawlow, who worked at Bell Labs.

In 1951 Townes left Columbia University and went to Bell Labs in New Jersey, and for the next few years he worked with Schawlow on the laser. By 1958 they had a design that they were satisfied with, and they published a

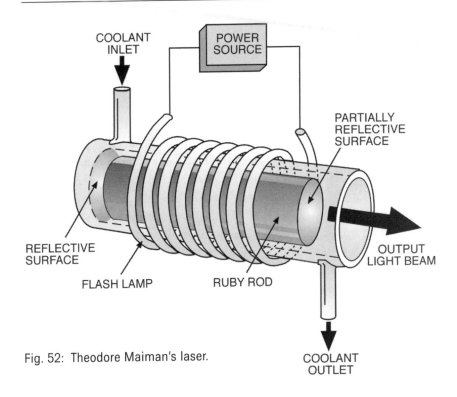

COOLANT
INLET

POWER
SOURCE

PARTIALLY
REFLECTIVE
SURFACE

REFLECTIVE
SURFACE

OUTPUT
LIGHT BEAM

FLASH LAMP

RUBY ROD

Fig. 52: Theodore Maiman's laser.

COOLANT
OUTLET

paper in *Physical Review.* Interestingly, about the same time, Gordon Gould, a graduate student at Columbia University, came up with the same design, but Townes and Schawlow had already patented their device.

Although they had the design, Townes and Schawlow did not build a working laser. The honor of building the first laser belongs to Theodore Maiman of Hughes Research Labs in California. It was built in 1960. Maiman used a ruby cylinder as the central piece of his device. One of the ends of the cylinder was a reflecting mirror, the other a partially reflecting mirror (see figure 52).

The light from a laser is coherent; in other words, the wavelengths are all the same, and they are all in phase (or lined up exactly). Because of this the beam can be sharply focused, and, as is well known, there are now numerous applications of the laser. They are used extensively in medicine, particularly surgery. Dermatologists, for example, use them routinely in the removal of skin legions. Lasers are also used in communication, surveying, weaponry, and many other places.

EINSTEIN'S PHOTONS

It is frequently said that many of Einstein's minor contributions to science could have won him the Nobel Prize, and indeed one did. Although it is not considered nearly as important as his relativity theories, his first contribution to quantum theory was a significant breakthrough. Planck had put forward his explanation of the heat curve in 1900, using the concept of quanta, but few took the idea seriously because it was so different. Einstein, however, did. In fact, as we saw earlier, he extended it to light, postulating that light and other radiations were made up of discrete particles of energy, which were later called photons. This was in conflict with Maxwell's theory, which treated radiation as electromagnetic waves. Einstein went on to show how his idea could be used to explain a phenomenon called the photoelectric effect. The German physicist Philipp Lenard had discovered that when light was shone on a metal surface, electrons were emitted. Strangely, the energy of the released electrons was independent of the intensity of the radiation; it depended only on its frequency. Furthermore, there was a cutoff frequency below which no electrons were ejected.

Einstein made the assumption that the electrons were absorbing the photons and gaining their energy. The energy depended on the frequency,

Fig. 53: The photoelectric effect.

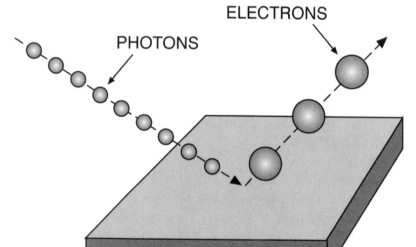

and if this energy was high enough, the electron would be ejected from the metal. Einstein wrote up the paper in March 1905 and submitted it to *Annalan der Physik*.[12] It was this discovery that eventually got him the Nobel Prize. Although the explanation of the photoelectric effect is, perhaps, one of his minor contributions, his hypothesis of the photon is certainly a major one. The understanding of photons has led to many modern devices. Among them are x-ray machines, CAT scanners, microwave ovens, and television screens, to mention only a few. All of these can be thought of as part of Einstein's legacy to us.

ATOMS AND MOLECULES

In early 1905 Einstein had still not obtained his doctorate. He had submitted several dissertations, but all of them had been rejected. Interestingly, he had even submitted his special theory of relativity, but it had been regarded as incomprehensible (actually, it was not complete at this time). Because of these difficulties, Einstein finally decided to stay away from anything controversial. He considered the area of atoms and molecules and decided it met the requirement. As strange as it might seem, several well-known physicists still did not believe that atoms and molecules existed. Einstein decided to prove theoretically that they did.

The breakthrough came one day while he was having tea with his friend Michelle Besso.[13] After putting sugar in his tea, Einstein realized that the viscosity had to change. Thinking about it for a few moments, he realized he might be able to use this change in viscosity to determine the size of the sugar molecule. He thought about the problem that evening and made a few calculations. Soon he had the answer. All he had to do was look up a couple of constants in a handbook at work the next day, and he would have the diameter of the sugar molecule. It turned out to have a diameter of one ten-millionth of a centimeter.

Einstein wrote up the work into a dissertation over the next few days, and a few weeks later he had his doctorate. This was an important breakthrough in that it was one of the first calculations of the size of a molecule, and it eventually helped convince scientists of their existence.

BROWNIAN MOTION

Einstein was pleased with his calculation of the size of a molecule, but he wanted to go further. Molecules at room temperature obviously had considerable energy and motion. Was this motion great enough to move tiny specks of matter—in particular, specks that we might be able to see in a microscope? Einstein explained his idea to Besso, and Besso remembered that such a phenomenon had already been discovered. About seventy-five years earlier, the Scottish botanist Robert Brown had observed grains of pollen floating on water using a microscope. He noticed that they had a jiggly motion, but no one had ever been able to explain the phenomenon.[14]

Einstein decided to work on the problem. Over the next few weeks he made several calculations; he visualized large spheres being knocked around by vibrating water molecules. All he could calculate was the average displacement of a given sphere, but that was enough. He showed that a tiny sphere of diameter one-thousandth of a millimeter would move about one-thousandth of a millimeter each second, and within a few years this was verified observationally. A few years later Einstein was nominated for the Nobel Prize for this discovery, but he did not receive it. As we saw, however, he did receive it a few years later for another of his "minor" discoveries, which eventually turned out to be of tremendous significance.

Chapter 10

Dreams of a Unified Theory

Einstein's goal for the last thirty years of his life was the formulation of a unified field theory. After the success of general relativity, it seemed natural to try to expand it to cover other fields and properties of nature. General relativity explained the gravitational field and how the universe was put together, but there was another field that operated at a different level. The electromagnetic field governed the world of charged particles and how they interacted. James Clerk Maxwell's theory explained this field, but Einstein soon became convinced that it could be incorporated in his general theory of relativity. After all, the two fields had many similarities. Matter was the source of both, although in the case of the electromagnetic field the matter had to be charged. Both fields had an infinite range, and both fell off in strength with distance; in other words, the farther you got away from the source, the weaker the field. Despite these similarities, however, there were significant differences. The electromagnetic field was 10^{37} times as strong as the gravitational field. Furthermore, the electromagnetic field had two sources, namely, positive and negative charges, and, as a result, there was both attraction and repulsion. The gravitational field, on the other hand, only attracted. Another interesting difference was that the electromagnetic field could be shielded; in essence, you could stop it. But nothing stopped the gravitational field.

Einstein, of course, understood all this, but it did not deter him. He was convinced that at some level the two fields were related. After all, it had been shown that the electric and magnetic fields were related, and at first glance they appeared to be quite different. But Einstein also had other problems to cope with. He wanted matter to appear naturally in his theory; in particular, he wanted it to predict the charge and mass of the known elementary particles. At the time only two of these particles were known: the electron and the proton.

THE FIRST UNIFIED FIELD THEORIES

Even though Einstein was thinking about how he could extend his theory, he was not the first to publish a unified field theory. Even before he published his general theory of relativity, an attempt was made to explain the existence of matter. Gustav Mie of Griefswald, Germany, used special relativity as the basis of a theory in which he attempted to explain the existence of the electron and the proton, but it was well before its time and was not successful.[1] A few years later David Hilbert of Göttingen incorporated the theory into general relativity, but it didn't help. It still did not explain the proton and the electron.

The first person to generalize Einstein's theory in an attempt to incorporate the electromagnetic field was Herman Weyl of Zurich, Switzerland. Weyl wrote one of the first comprehensive texts on general relativity. Titled *Space, Time, and Matter*, it was so complete and elegant that it amazed scientists and thoroughly impressed Einstein. Shortly after he published the book, Weyl extended general relativity by assuming that length and direction were not preserved as you moved in curved space.[2] He sent his theory to Einstein, asking him to present it to the Prussian Academy and publish it in the academy *Proceedings*. Einstein was impressed with the theory, but after studying it in detail, he found a flaw. If distances changed, time also had to change, and this meant that the history of a particle affected its properties. In short, if particles took different paths to the same point, they would end up with different vibrational frequencies, and it was well known that this did not happen in nature. Despite this short-coming, Einstein sent it in for publication, but he added a comment at the end of the paper, pointing out the problem he had discovered.

Einstein was now even more convinced that the gravitational and electromagnetic fields could be brought together. Even though Weyl's idea was wrong, it was ingenious. It was, in fact, later used in quantum theory. Einstein was certain that there were other equally ingenious approaches, and one of them would have to work. But before he could formulate his own theory, another one arrived in the mail. Theodor Kaluza of Königsberg, Germany, came up with another ingenious way to extend general relativity.[3] He wrote down the equations of general relativity in five dimensions rather than four, and to his delight, he discovered that the additional equations in the new theory were Maxwell's equations of the electromagnetic field. Again it appeared as if the gravitational and electromagnetic fields had been brought together.

Einstein was amazed at Kaluza's approach, and again he was sure it was the big breakthrough. But he soon discovered that the theory had problems. First of all, it didn't predict the mass and charge of the electron and the proton. Also, there was the meaning of the additional dimension. We only observe four dimensions in our world. What did the additional dimension represent physically? Over the next few years Einstein's enthusiasm waned, and he eventually lost interest in the theory. In 1926, however, Oscar Klein of Sweden gave a reasonable explanation of the extra dimension. He said it was curled up and was so small that it was unobservable. Today, many of our theories have extra dimensions, and this is still the explanation we use to justify the unseen dimensions.

By the time Klein explained the extra dimension, another problem was arising for Einstein. He still hoped to explain the microworld using his new theory, but an entirely new and different theory had been put forward, namely, quantum mechanics, and it was extremely successful. The new theory was based on the concepts of probability and uncertainty, concepts that were foreign to general relativity. If Einstein was to come up with a unified field theory by extending general relativity, he would have to explain quantum mechanics. In short, he would have to show that quantum mechanics was contained within the new theory.

Finally, in 1925 Einstein devised a theory of his own, and like the earlier attempts, it was ingenious. He asked himself what the simplest and most natural extension of general relativity was, and soon found an approach that had not been tried. General relativity was a symmetric theory; in other words, the equations were symmetric in the same way a sphere is symmetric. If you draw a line down the middle of a sphere, the two sides are the same. Einstein decided to give the theory a nonsymmetric part, and when he did, he found to his delight that the extra equations that came out of the theory were Maxwell's equations of the electromagnetic field.[4] Einstein was excited by the result. He wrote to his friend Besso expressing his enthusiasm for the new theory, telling him he was sure he had finally found what he was looking for. But when he examined the theory in detail, he found that it, too, had problems. It did not predict the mass and charge of the electron and the proton; furthermore, it predicted something that confused him. It appeared to predict mirror-image particles, in other words, particles of the same mass but opposite charge. For example, corresponding to the electron there was a particle of the same mass but opposite (positive) charge. Since no particles of this type had been found in nature, Einstein saw this as a flaw in the theory. As in the case of the cosmological constant, where he missed an opportunity to predict the expansion of the universe, he missed another opportunity.

A few years after Einstein found these strange particles, Paul Dirac of Cambridge University used a different theory to predict the same particles. And they were soon discovered. We now refer to them as antiparticles.

Einstein spent the rest of his life searching for a unified field theory. Time after time he would discover a new theory and think he had finally found the "true theory." Then he would find a flaw and discard it. After all his success, it was a frustrating process for him, but he never gave up. Even to the last days of his life he continued his search. Many considered him a little crazy—chasing after a rainbow. But eventually a number of scientists began to realize he was merely ahead of his time, and they began to follow his lead.

THE PROBLEM

The major difficulty with Einstein's quest was that it was too restricted. When he began, it was a reasonable search: only two fields, namely, the gravitational and electromagnetic fields, were known, and two particles, the electron and the proton. But over the years more fields were discovered, and hundreds of new particles were found. Aside from this, though, there was another problem. When Einstein was working on his earlier theories, particularly his special theory of relativity, he had a slight disdain for mathematics, thinking of it only as a tool. He concentrated on physical principles and was sure that new physical insights were needed. With the mathematical complexity of general relativity, however, he began to look upon mathematics as more than just a tool. Rather than looking for new physical insights, he began to play with the mathematics, searching for an equation, or set of equations, that explained things. He began to rely more and more on this method, and it eventually created serious problems for him.[5]

Getting back to the earlier difficulty, however, it's easy to see why Einstein's approach was shortsighted, although he no doubt hoped the new theory would take care of the things he was ignoring. First of all, while he was searching for a unified theory, two more fields were discovered: the strong and the weak nuclear forces. The strong force holds the particles of the nucleus together, and the weak force is important in certain types of nuclear reactions. In addition, the array of newly discovered particles soon ran into the hundreds. They became so numerous, in fact, that physicists soon began to realize that they all couldn't be elementary. Most of them had to be made up of "more elementary" particles.

The heavy particles were called *hadrons,* the lighter ones, *leptons* (within the hadrons there were subclasses of baryons and mesons). There were so many of them that physicists soon realized that a new approach was needed. And in the early 1960s Murray Gell-Mann and George Zweig, both of Caltech, independently suggested that hadrons were made up of particles that Gell-Mann called *quarks.*[6] Originally, there were three of these quarks, referred to as up, down, and strange. The hadrons were composed of various combinations of them. For example, the proton was composed of two up and one down quark. Eventually, it was shown that there were three more quarks; they were referred to as charmed, bottom, and top. But these quarks had never been seen, and this had to be explained. Gell-Mann suggested that they were hidden in a "bag" and could not get out of it, and the idea was soon accepted.

Einstein's simple union of the gravitational and electromagnetic fields obviously could not account for all of this. Furthermore, quantum

Fig. 54: Quarks
trapped in a "bag."

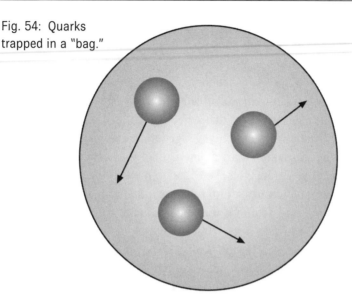

mechanics was soon explaining many of the interactions between particles and radiation. To see how it did this, we have to go back to Heisenberg's uncertainty principle. It tells us that we cannot measure certain
variables simultaneously to a high accuracy. For example, if you focus in
on the momentum of a particle and measure it accurately, the particle's
position becomes fuzzy. And if you try to measure its position accurately,
its momentum becomes fuzzy. The same thing applies to energy and time.

Because of this fuzziness, particles can "loan" energy for a very short
period of time. The only stipulation is that they pay it back before the
"cloak of fuzziness" disappears. This means that a particle and its antiparticle (e.g., an electron and a positron) can form briefly in space. We
cannot observe them directly, however, because the loan of energy that
generated them has to be paid back quickly. Nevertheless, these particles
and antiparticles do exist, and we see evidence of them in many experiments. They are referred to as *virtual particles*.

The particle and antiparticle that are created quickly annihilate one
another, and in the process photons are released. Because of this, we can
think of each particle in space as surrounded by a "cloud" of photons.
When another particle passes nearby, some of these photons are
exchanged between the two particles. If both particles are positively
charged, the exchanged photons will cause a repulsion between them. If

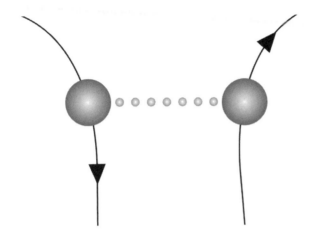

Fig. 55: Particles passing close to one another.
They are interacting by passing photons back and forth.

one is positively charged and one negatively charged, on the other hand, the photons will cause an attraction. From a quantum point of view, therefore, the electromagnetic force between two particles is seen as the exchange of photons. Physicists developed this idea into what is called quantum electrodynamics (QED), and it soon became the most accurate and successful theory ever devised.[7]

Encouraged by the success of QED, physicists Sheldon Glashow, Abdus Salam, and Stephen Weinberg applied the same approach to the strong and weak nuclear reactions. In the case of the strong interactions, particles called *gluons* were assumed to be exchanged between the quarks making up the particle. The theory was called quantum chromodynamics (QCD) because the quarks and gluons were assumed to be colored (color in this case, however, was something different from its usual meaning; it was merely a label, or attribute, of the particle). Again, the theory was extremely successful. A similar theory was developed for the electroweak interactions. In this case the exchange particles were weak bosons called W and Z.[8]

With the success of quantum chromodynamics and quantum electroweak theory, it was natural to try to bring them together into a unified theory. This attempt was called grand unified theory, or GUTs, for short.

It has been quite successful, but a few problems remain. Nevertheless, of the four known forces of nature, we can say with some confidence that physicists have unified three of them. One, however, namely, the gravitational field, remains outside the fold. Many attempts have been made to bring gravitation into the fold in the same way the other fields were incorporated. In these attempts, exchange particles called *gravitons* are assumed to exist, and when two masses pass close to one another, they exchange gravitons. To do this mathematically, however, physicists have to "quantize" general relativity, and that has proven to be difficult. Despite many attempts, the theory has not been quantized.

It appeared as if the two great pillars of modern physics, namely, general relativity and quantum theory, were incompatible. To see why, we must go back to the uncertainty principle. As we saw earlier, something strange happens when we narrow in on a very small region of space. According to classical theory, empty space has no gravity, but quantum mechanics tells us that this isn't the whole story. Because of the uncertainty principle, tiny undulations exist on a small scale. To see the significance of this, let's assume we zero in on a small section of space and magnify it to higher and higher power. At first the space is flat as predicted by general relativity, but as we get to higher magnifications, we finally begin to see undulations, and as we continue zeroing in, space becomes more and more frothy and twisted. John Wheeler called this the "quantum froth." But we know that general relativity can't deal with it; it recognizes only flat, smooth space, so there is obviously a conflict.

We can easily determine where this frothiness begins. Using the basic constant of quantum theory, namely, Planck's constant, and the universal gravitational constant, we find that it begins at what is called the Planck length, which is 10^{-33} centimeter. It goes without saying that this is a very small distance; it's twenty orders of magnitude smaller than the atomic nucleus. Indeed, it is so small that it might seem that it is of little significance. But it is important for two reasons. First of all, it seems inconceivable that there would be two distinctly different theories governing the universe: one for the very small and one for the large. And second, there are phenomena that cannot be explained without it.

For most phenomena in our universe, the two theories do not overlap; in other words, we need one or the other, but not both. There are cases, however, where both are needed. One is the center of black holes, and the

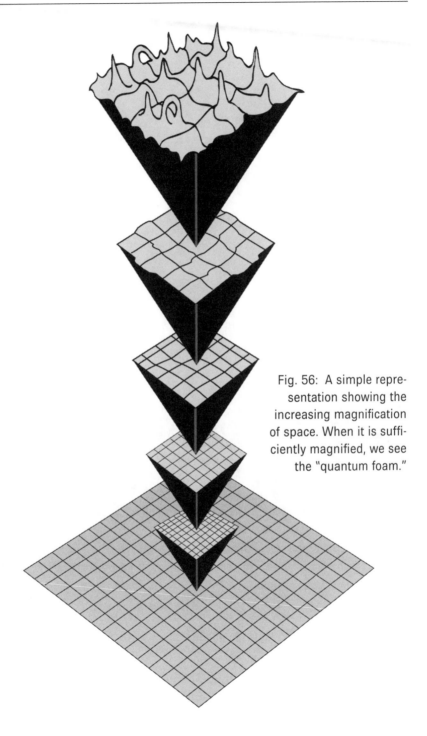

Fig. 56: A simple representation showing the increasing magnification of space. When it is sufficiently magnified, we see the "quantum foam."

Fig. 57: A closeup of the quantum foam.

other is the early universe. We can't understand the earliest stages of the big bang without a union of quantum theory and general relativity. We refer to this union as quantum gravity.

THREE ROUTES TO QUANTUM GRAVITY

As we saw earlier, Einstein struggled to find a unified theory for years and never succeeded. During most of this time, his endeavor was frowned upon by most scientists. But eventually a few began to realize it was an important problem, and they began their own research programs. Some used the same approach he did, namely, they worked mainly with general relativity, trying to generalize it. Most, however, worked with quantum theory, or, more exactly, grand unified theory, trying to extend it to include gravity.

For the most part, physicists are now approaching quantum gravity from three fronts.[9] Since we're trying to bring general relativity and quantum theory together, one path has to start with general relativity, and the other with quantum mechanics. The path from general relativity is the one Einstein followed. In this approach, the hope is to generalize or extend general relativity so that it includes quantum theory and explains the elementary particles and so on. With Einstein's lack of success, it might seem that there would be little hope with this approach, but recently a significant breakthrough has been made. A reformulation of the equations of general relativity has led to what is called *loop quantum gravity*, and over the last few years it has shown considerable promise. Several important breakthroughs have been made using it.

The second approach is through quantum theory, and over the past few decades much of the work has been in this area. The most promising approach is what is called *string theory*, and most people are working in this area. A number of important developments have been made in it in recent years. (They will be discussed in the next chapter.)

Most scientists working on quantum gravity are engaged in one of the two areas discussed above, but a number of people are taking a third approach. They feel that both of the above approaches are doomed to failure and that an entirely new approach is needed. They are taking a fresh look at space and time and trying several innovative ideas. Twistor theory, in which particles and space are assumed to be made up of small "twisters," is an example of such an approach.

But even if a union of relativity and grand unified theory succeeded, another problem would remain. It would be important only at distances that were so far removed from our experiences that no experimental tests of it would ever be possible. At least that is what was believed for many years. Recently, though, there has been an important breakthrough in this area. A number of experiments have been suggested that might make quantum gravity testable. The idea is to use the universe—cosmic rays, the cosmic background radiation, and gamma-ray bursts from black hole and neutron star collisions—rather than accelerators to test the theory. (At the present time, accelerators are not capable of testing the theory.)

LOOP QUANTUM GRAVITY

The breakthrough to what is now called loop quantum gravity came in the early 1980s, when a postdoctoral student at the University of Maryland, Amitaba Sen, made an important discovery while attempting to quantize a theory called supergravity. Abhey Ashtekar, now at Penn State, realized that Sen's breakthrough would allow a reformulation of general relativity, and over the next few months, he set up the new theory and found that it was ideally suited for quantum calculations.[10] In fact, it gave rise to an equation that had been written down in the 1950s by John Wheeler and Bryce DeWitt, an equation that no one had ever solved. With the reformulation, however, the Wheeler-DeWitt equation was simpler, and a solution was soon obtained. The solutions are in the form of "loops," and this is why the new theory is called loop quantum gravity. Quantum states of the geometry of space are expressed in terms of these loops.

One of the important results of loop quantum gravity was that it showed that space at the scale of the Planck length was discontinuous. It was made up of discrete units; in other words, it was quantized. The minimum allowed volume is the cube of the Planck length, and all other volumes are numerical multiples of it.

An important advance in the theory came when several scientists realized that a discovery made several years earlier by Kenneth Wilson might be helpful in visualizing the results.[11] Wilson had discovered a lattice that could be used to exhibit the properties of quantum chromodynamics. Lee Smolin, who was then at Penn State, and others showed that this lattice gave a convenient way of exhibiting the solutions in loop quantum gravity. It shows, in effect, that space is constructed from the relationships between a set of discrete elementary loops. We now take the point of view that these loops do not live on the lattice, or even in space, rather, their interactions define the space. Indeed, we can say that quantum gravity is, for the most part, the theory of the interacting, knotting, and linking of loops.

Recently, scientists have discovered that loop quantum gravity is in conflict with special relativity in that it implies that the speed of light is not necessarily a constant. According to loop quantum gravity, it has a small dependence on the energy of the field: photons of higher energy travel at slightly slower speeds than those of lower energy. Scientists may be able to check this out in 2006 when the satellite GLAST (Gamma-ray Large Area Space Telescope) is launched. One of the most interesting consequences of this variation comes from cosmology. As we saw earlier, a cosmology has been put forward that explains inflation and several other aspects of the early universe in a very simple way if the velocity of light was slightly higher in the early universe. Loop quantum gravity may therefore be giving us new insights that will help us understand our universe better.

THE BLACK HOLE CONNECTION

Another important development in loop quantum gravity is that it has been shown to be closely linked to black holes.[12] We discussed black holes earlier and showed that they have many intriguing properties. One of them is that the interior of a black hole is hidden behind an event horizon. This event horizon is helpful to us in that it acts like a microscope that enlarges space; it allows us to see the Planck length directly. It does this because of an effect we discussed earlier. We noted that as an astronaut fell into a black hole, his clock appeared to go slower and slower as seen by an outside observer. It almost stops as it approaches the event horizon, but it never quite stops because it never quite reaches the event horizon. This means that, for an emerging photon, time is stretched out, and this, of course, changes its frequency, or wavelength. In other words, the horizon of a black hole stretches out the wavelength of an emitted light beam. Therefore, if we look at the light coming from close to the horizon, we are able to see the quantum structure of space.

Interestingly, another event horizon has been discovered that is quite similar to the event horizon of a black hole. Furthermore, even though it has nothing to do with black holes, it has many of the same properties. It was discovered by William Unruh of the University of British Columbia.[13] To see how it arises, consider an observer moving uniformly

Fig. 58: Left: Spaceship is not accelerating.
Right: Spaceship is accelerating and a "horizon" forms.

through space. As you would expect, he sees nothing but empty space around him. Now assume he begins to accelerate. Unruh showed that the observer will now see the space around him filled with particles. Where do these particles come from? Earlier we saw that space is filled with virtual particles because of the uncertainty principle; it is these particles that he is seeing. Furthermore, Unruh demonstrated that these particles have a temperature, and the faster the observer accelerates, the higher the temperature. Since the particles are virtual, they annihilate rapidly to photons, and because of this, our observer will find himself embedded in a hot gas of photons and particles that form a horizon that he cannot see through. Of course, if he suddenly stops accelerating, the horizon disappears. This is not the case with a black hole horizon, so the two types of horizons are different. Nevertheless, there are many similarities. For example, both conceal entropy.

Incidentally, because of this effect, we can say that Einstein's principle of equivalence has to be generalized. It predicts that an accelerating observer experiences a gravitational field. We now have to generalize it to say he will also experience a horizon.

To appreciate the importance of this, we must first look at some of the other developments in black hole physics. We saw earlier that entropy plays an important role in relation to black holes. Entropy is a measure of how much disorder there is in a system. Furthermore, in 1972 Jacob Bekenstein, who is now at the Hebrew University in Jerusalem, showed that the entropy contained in a black hole is proportional to its surface. In addition, shortly thereafter Stephen Hawking of Cambridge University showed that black holes emitted radiation, and since this radiation carried off energy, and since energy and mass are equivalent, the black hole's mass decreased and it shrank slightly. Furthermore, as it lost mass, it got hotter. Indeed, it kept shrinking as it radiated, and it continued to get hotter and hotter until it finally exploded. There is a problem, however, with this assumption. We know that entropy is a measure of information, and if the black hole explodes, this information is lost. This means the black hole violates the second law of thermodynamics. Scientists obviously had to do something about this, so they revised the law.

The revised second law of thermodynamics says that the entropy of the black hole plus the entropy of the region around it cannot decrease. To see the significance of it, let's go back to Hawking's prediction of evaporation. Hawking explained that this evaporation is due to the severe twisting and distortion of space near the event horizon. Because of it, the virtual particles that are created in its vicinity become separated, and some of them escape from the black hole. We can also look at this from a different point of view. Assume a black hole swallows an atom or photon. It obviously gains energy, or mass, and therefore its entropy increases. But if the entropy inside the black hole increases, the entropy outside it must decrease. We can, of course, look at this from the other direction. Assume the horizon shrinks; its entropy therefore decreases, and the entropy outside the black hole has to increase. But if it increases, particles—photons—must be created outside the black hole. This is the Hawking radiation we observe.

The importance of all this to loop quantum gravity is that it is all predicted by loop quantum gravity.[14] Loop quantum gravity predicts a discrete, quantized structure of space, and it can also explain the entropy and temperature of a black hole. Furthermore, Unruh's discovery of a horizon for an accelerated observer gives us a second way of looking at the properties of space near a horizon, and has proven to be invaluable. Unruh's law coupled with Bekenstein's law of entropy has given us a valuable tool for looking at black holes and their link to quantum gravity.

Fig. 59: Hologram of a woman's face.

THE HOLOGRAM CONNECTION

As just mentioned, entropy is associated with information. The surface area of a black hole is a measure of this entropy, and if it has entropy, it also contains information. What exactly is information? In this era of computers, CDs, and so on, it is a familiar concept. It's what is stored on a hard drive, or on a CD. It's what gives us the instructions to build something. A black hole contains information, but because of the event horizon, it's impossible for us to get at it. No information can pass through an event horizon. As it turns out, though, we can still learn something about it. The generalized second law of thermodynamics allows us to set bounds on the information capacity of any isolated physical system. Bekenstein put forward what he called the universal entropy bound (we now refer to it as the Bekenstein bound), which limits how much entropy and therefore how much information can be extracted from a black hole; according to his hypothesis, it is proportional to the area of the surface.[15] In 1995 Leonard Susskind of Stanford University devised a similar bound, called the holographic bound. It limits how much entropy can be contained in the matter and energy occupying a certain volume of space.

Since holograms play a vital role in what follows, I will explain briefly what they are. A hologram is a special photo that generates a three-dimensional image when it is illuminated in the proper way. In effect, all the information needed to form the three-dimensional image is coded on the two-dimensional surface of the photo. Bekenstein has pointed out that we can apply the holographic principle to the universe. He has shown that the four dimensions of the universe (three of space and one of time) can be coded in three dimensions, and in this respect it is like a hologram. Actually, there's a little more to it than this. The universe is actually assumed to be a "network" of holograms, each of which contains information about relationships. There is still some controversy about this idea, but it has shown a lot of promise.

EINSTEIN AND THE LATEST DEVELOPMENTS

Einstein didn't succeed in his quest to find a theory of everything, but he opened the door and showed the way. For many years most scientists ignored his work, but after he died, many people followed his lead, and in recent years considerable progress has been made. Loop quantum gravity and string theory are both relatively successful and have made a number of interesting predictions. But we now have a problem, much like the one that arose after two successful formulations of quantum mechanics were made in the mid-1920s, one by Heisenberg and one by Schrödinger. Are loop quantum gravity and string theory distinct and incompatible? Or are they, as Schrödinger showed regarding the two approaches to quantum mechanics, just two different formulations of the same theory?

Chapter 11

Strings and Superstrings

Einstein's dream may be realized by one or the other, or possibly both, of the two theories mentioned in the last chapter: loop quantum gravity and string theory. Loop quantum gravity is an extension of general relativity and follows the route Einstein took for many years. String theory, on the other hand, begins with quantum theory and attempts to incorporate general relativity. Over the past few decades it has generated a lot of interest, and scientists have developed it extensively. In this chapter we will take a closer look at these developments.

THE BEGINNINGS

String theory developed from something that appeared to have no connection with strings. In 1968 Gabriele Veneziano, now of CERN, published a paper on what he called the dual-resonance model of strong interactions.[1] He hoped to explain some of the problems of the strong interactions using his theory, but he was thinking strictly in terms of point particles. Two years later, in 1970, Yoichiro Nambu of the University of Chicago, Leonard Susskind of Stanford University, and Holger Nielson of the Niels Bohr Institute in Copenhagen independently noticed that there was a relationship between dual resonance theory and the vibrational states of a string. As you no doubt know, a string of fixed length that is tied down at each end can vibrate with one, two, three, or more loops along its length. We refer to each of these different modes of vibration as a resonance. Nambu, Susskind, and Neilson noticed that these resonant states could be related to the known elementary particles. In short, the elementary particles could be described by the resonant states of a string.

One way of looking at these states is to compare them to the vibrations of a guitar string. The various notes of the musical scale can be sounded by plucking a guitar string in just the right way. Indeed, a large number of different notes (higher scales) can be sounded if the strings are tight enough. This means that a large number of elementary particles could be accounted for using the modes of a vibrating string.

What did these strings look like? First of all, they were all the same; it was only the vibrational modes that differed. And they were assumed to be about the size of a proton, or about 10^{-13} centimeter long. They had no mass and were elastic. But if they had no mass, you immediately wonder how they accounted for the particle's mass. As it turns out, it was related to the tension of the string. Finally, there were both open and closed strings; the closed strings took the form of loops.

The theory was interesting, but it had problems. The major one was that the elementary particles were known to be made up of two classes, referred to as *bosons* and *fermions*, and the fermions were not accounted for. Fermions are matter particles that have half-integral spin (the spin of an elementary particle is similar to the spin of a ball, but there are differences). The electron is an example of a fermion. The bosons have integral (1, 2, 3, 4 . . .) spin.

The theory could account only for bosons. Furthermore, it predicted a particle that was viewed by most physicists with skepticism. This particle traveled only at speeds greater than that of light and is referred to as a tachyon. No one had ever detected or seen a *tachyon*. Another difficulty was that it required a space of twenty-six dimensions, which was hard to explain.

The association between particles and strings was interesting, but it

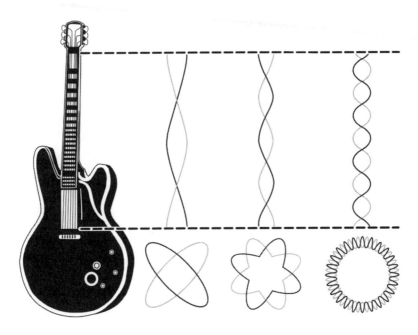

Fig. 60: The nodes and loops formed by a guitar string.
They are similar to those used in string theory.

appeared to have too many difficulties. Then in 1971 Pierre Raymond,
now at the University of Florida, overcame what appeared to be the
greatest of them.[2] He showed that the theory could be extended to include
fermions. André Neveu of France and John Schwarz, now of Caltech,
improved the theory and reduced the number of dimensions to ten. But
the tachyons were still there, and the theory contained other particles that
didn't exist in nature.

Most scientists still considered particles to be points, and much
progress was being made on this front. Quantum chromodynamics, which
was formulated in the mid-1960s, appeared to solve many of the problems related to the strong interactions. As a result, few people were interested in pursuing strings. But John Schwarz had faith in the idea and continued working on it.

THE STANDARD MODEL

Over the next few years, what is now called the standard model was developed, and its predictions agreed well with experiment. Furthermore, it seemed to explain the interactions associated with the electromagnetic, strong, and weak interactions. String theory obviously had serious competition. In the standard model, elementary particles are grouped into two classes according to their spin: fermions and bosons. Fermions are the matter particles, and bosons are the particles that transmit the forces observed in nature (they are referred to as "exchange particles"). The bosons are as follows:

> photon—exchange particle of the electromagnetic force
> gluon—exchange particle of the strong interactions
> W^+, W^- , Z^0—exchange particles of the weak interactions
> graviton—exchange particle of the gravitational field

(Technically, the graviton is not part of the standard model, but I have included it for completeness.)

The fermions appeared to be grouped into three generations, which are as follows:

Electric charge

electron	−1
muon	−1
tau	−1

Each of the above particles has a neutrino as follows:

Charge

electron neutrino	0
muon neutrino	0
tau neutrino	0

There are a total of six quarks grouped into two families. Quarks make up the hadrons.

Charge

up	2/3
charmed	2/3
top	2/3

And

down	−1/3
strange	−1/3
bottom	−1/3

It's important to note that the mass increases as we move downward in each list. One of the unsolved problems of the standard model is why there are three families in each group. The biggest shortcoming of the standard model, however, is that, although it explains three of the known forces of nature very well, it says nothing about the fourth, namely, gravity. Numerous attempts have been made to bring gravity into the fold, but none have succeeded. The hope is that string theory will be able to do this.

THE FIRST STRING QUANTUM GRAVITY THEORY

One of the major embarrassments of string theory in the 1970s was that several particles were predicted that seemed to have nothing to do with the standard model. One of these particles was massless. In 1974 Joel Scherk of the École Normale Superiéure and John Schwarz of the California Institute of Technology began looking carefully at this particle, and they finally realized it had all the properties of the graviton—the hypothetical exchange particle of the gravitational field.[3] Schwarz and Scherk were delighted. If string theory predicted the graviton, it might be even more powerful than they had hoped. It might be a quantum theory that included gravity. Indeed, it may even be a theory of everything (TOE). They wrote up a paper announcing their results in 1974, suggesting that string theory may be an all-encompassing theory.

You might imagine that such an announcement would cause a sensation. It didn't. In fact, hardly anyone paid any attention to it. Most physicists were still viewing string theory with skepticism. In 1976 Scherk, along with Ferdinando Gliozzi of the University of Turin and David Olive of Imperial College of London, published another paper showing that it might be possible to incorporate supergravity into string theory. Supergravity was an extension of general relativity that had been devised a few years earlier in which a certain type of symmetry had been added. But again few people paid any attention.

THE SCHWARZ-GREEN BREAKTHROUGH

The prospects for string theory seemed dismal, but Schwarz was confident and continued to work diligently on the theory. For several years, however, he published nothing. Then in the summer of 1979 he met Michael Green of Queen Mary College while both men were at CERN. Green was also working on string theory, and the two men soon decided to work together. Scherk and several of his colleagues had tried to show that string theory, was supersymmetric. Supersymmetry is a symmetry in which bosons and fermions are two states of the same particle. In addition, they wanted to prove that the resulting theory was free of infinities. (When calculations were made in most theories that attempted to incorporate gravity, the answers always came up as infinite, showing that the theory was flawed.) The two men worked on the theory for the rest of the summer, but they came up with nothing. They agreed to meet the following summer.[4]

During the summer of 1980 they were together again at Aspen, Colorado, and this time their work paid off. After several months of tremendous effort, they were able to develop a string theory that contained supersymmetry. They published their results in 1981. They were sure it would attract a lot of attention. After all, they now had a theory that contained no infinities, explained all four forces of nature, including gravity, and appeared to predict the elementary particles. But again it went almost unnoticed.

The theory still had a problem. Schwartz and Green had not been able to show that the theory was "anomaly-free." Anomalies are things that create crazy results, such as negative probabilities. Over the next two years they worked to find a version of the theory that was anomaly-free.

What they wanted was a group theory (a mathematical theory that concerns itself with elements of a group and how they change under various operations) that could be used as the basis of their theory, and make it anomaly-free. They tried all the different groups they could think of; finally, they found one that worked. It is referred to as SO(32). They were pleased but continued their search, and found another, even better one. It is referred to as $E_8 \times E_8$. They finally had what they had been looking for: an infinity-free theory that explained all four forces of nature and the elementary particles—and was anomaly-free to boot.

They made their announcement in August 1984, and this time the scientific world finally took notice. One of those who became particularly excited was Edward Witten of Princeton University.[5] Using an entirely different approach, he showed that the theory was, indeed, anomaly-free, and he did it in record time. This convinced him that the theory was worth taking seriously. A few months later David Gross of Princeton University, Jefferey Harvey of the University of Chicago, and Emil Martinec and Ryan Rohm of Princeton University found another anomaly-free theory, which they referred to as the "heterotic" model. It was a closed-string model.[6]

With the new theories came the realization that the size of the strings in the old theory was wrong. Previously, they had been assumed to be about 10^{-13} centimeter long—about the size of the proton. But when gravity was incorporated into the theory, their size decreased considerably. According to the new theory, they were of the order of the Planck length, which is 10^{-33} centimeter. This is one hundred billion billion times smaller than the nucleus.

THE FIRST STRING REVOLUTION

The years between 1984 and 1986 were soon referred to as the first string revolution.[7] With Schwarz and Green's breakthrough, physicists around the world jumped on the bandwagon, and for a few years there was a tremendous amount of enthusiasm. Within the next few years more than a thousand papers were written on the subject worldwide, and a number of important advances were made. But problems persisted. One of the major ones was writing the equations of the theory in an exact form, or at least in a very accurate form. Because of this, approximate equations were used, and even they were difficult to solve. The solutions were therefore only approximate, and, in many cases, the equations were impossible to solve exactly. Nevertheless, a lot was learned about the various vibrational modes of the strings and how they interacted.

With the mounting difficulties, enthusiasm for string theory began to wane, and many of the people working in the area began to look for more fertile fields. The late 1980s and early 1990s were lean times for the theory. Few advances were made, but the diehards stuck with it.

STRING INTERACTIONS

Particles interact, and we know from experiment what the consequences of these interactions are. We can predict the results of these interactions with the standard model. But if string theory is to include the standard model, we have to have a way of displaying and calculating these interactions. Let's begin with a review of the properties of strings. We know that both boson and fermion strings exist, with each vibrational state representing a particle. We also know that waves can run either clockwise or counterclockwise along the closed strings. This is helpful in that it gives an explanation of the some of the properties of particles associated with their spin. Furthermore, if we consider a given string over time, its vibrational state is stretched out into a "sheet." This sheet is called a "world sheet." For closed strings, this sheet is bent into a cylinder.

Of particular importance in relation to strings is the *tension* they are under.[8] The greater this tension, the greater the energy of the string. (More accurately, the energy depends on both the tension and the mode or type of vibration.) The fundamental tension is referred to as the Planck tension, and it is incredibly high—10^{39} tons. From it we can get what is called the Planck mass, which is 10^{19} times the mass of the proton. And since the mass equivalent of a vibrating loop is a whole number times this, it raises the question: how can such high energies (or masses) account for all the elementary particles? Most of them are much lighter than this. As it turns out, several of the terms in the calculations cancel one another, leaving us with masses in the required range.

Now we ask: how do we represent the interaction of particles using strings? In quantum electrodynamics and quantum chromodynamics, these interactions are represented by simple line drawings called Feynman diagrams. A line represents a particle moving through space. A typical interaction is shown below.

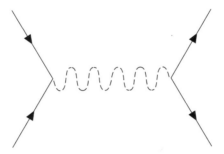

Fig. 61: A Feynman diagram.

This represents the collision of, say, an electron and a positron, giving rise to a photon, which later produces another electron-positron pair.

In string theory our particles are strings, so we have strings at each point of time as shown in figure 62.

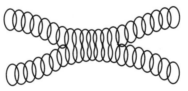

Fig. 62: A Feynman diagram in string theory.

The world sheet of this looks as follows.

Fig. 63: World sheet of the above diagram.

SUPERSYMMETRY

Supersymmetry plays such a large role in string theory, it's important to take time to consider it.[9] Although some early work had been done on supersymmetry, the first complete theory was formulated by Julius Wess of Karlsruhe University in Germany and Bruno Zumino of CERN, but it was not a string theory. They were concerned only with point particles. They worked together for several years at NYU, and it was during this time that the breakthrough in supersymmetry occurred. Wess and Zumino were interested in bringing the two types of particles in the universe, namely, fermions and bosons, together into the same family, and they were delighted when they found that by applying symmetry principles they could mathematically change a boson into a fermion and vice versa. We now refer to this as *supersymmetry*. It was particularly important because it seemed to imply that the universe was simpler than previously realized.

What they found was that supersymmetry theory predicted that every boson (particle that transmits forces) has a fermion (matter particle) partner. These partner particles are now referred to as *superpartners*. This means that for particles such as the electron, the quark, and the graviton, there are superpartners; we refer to them as the selectron, the squark, and the gravitino. A summary of the superpartners of the bosons with their names is as follows:

Particle	Superpartner
graviton	gravitino
photon	photino
gluon	gluino
W^+	Wino+
Z	Zino

And for the fermions:

Particle	Superpartner
electron	selectron
muon	smuon
tau	stau
neutrino	sneutrino
quark	squark

So far we have found no evidence of these superpartners. The most likely reason is that they are too heavy to be observed with our current accelerators. It is hoped that within the next few years, with larger, more energetic accelerators coming on line, they will be found.[10]

The addition of supersymmetry to string theory was a major breakthrough. The new theory became known as superstring theory, and with this new theory, several of the problems of string theory were solved. One of the most important was related to the strengths of the three forces of nature. As we saw earlier, Glashow, Salam, and Weinberg showed that the electromagnetic and weak forces were connected. Then in 1974 Glashow and Georgi showed that this union could be merged with the strong interactions to give a grand unified theory. In particular, they demonstrated that at very high energies, such as those that existed in the very early universe, these three forces were unified, in that they all had the same strength.

Let's look at this from a different, but equivalent, point of view. Assume we probe a particle such as an electron that exhibits the electromagnetic force. We know that there is a cloud of particles and antiparticles around the electron, and this cloud obscures the electron's true electric field. As we get closer to the electron, we penetrate this cloud and find that the electron's electric field increases. Eventually, we see the "true" electromagnetic force associated with the electron. But there are two other fields: the strong and the weak nuclear forces. What happens when we penetrate them? It turns out that they get weaker. Indeed, at very short distances, all three of the above forces are equal.

When these calculations were first performed, however, there seemed to be a problem. The strengths of the three nongravitational forces were approximately equal when probed at a sufficiently small scale, but they were not *exactly* equal. This didn't appear to make any sense. It seemed much more reasonable that they would be exactly equal. And, indeed, when supersymmetry was added to the theory, they became exactly equal.

But supersymmetry's greatest success is that it gave us a more satisfactory union of general relativity and quantum theory. The supersymmetric union avoids the problems of tachyons, and it appears to explain all the elementary particles of nature. Furthermore, it helps explain the major conflict between the two theories. As we saw earlier, this conflict centers on the quantum fluctuations that occur at the Planck scale. In effect, these fluctuations are so violent at this tiny scale that they destroy

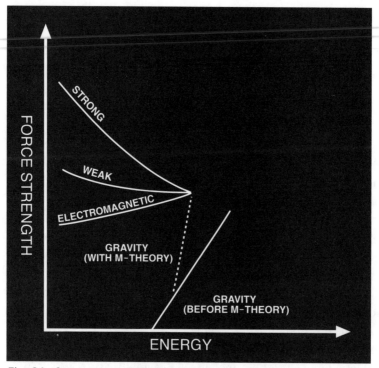

Fig. 64: Supersymmetry (and M-theory—see next chapter) helps bring gravity and other forces of nature together in the early universe.

the smoothly curving space that is needed in general relativity. Superstring theory smooths them out.

ADDING MORE DIMENSIONS

Let's turn now to the meaning of the higher dimensions of the theory.[11] Superstring theory is a ten-dimensional theory. How do we account for all these dimensions? Our world is made up of only four dimensions. Where are the extra dimensions? As we saw earlier, Klein and Kaluza assumed that they were curled up so small that they weren't observable, and, as you might expect, this is what is assumed in superstring theory. But even with this, there is still a problem. We can easily visualize the curling up of a single dimension; it would look like a tiny cylinder. But how do we visualize the curling up of several dimensions? Before I get into this, I'd like to remind you that this curling up occurs at every point in space, so it's difficult to represent in a diagram.

Let's begin with two curled-up dimensions. We can think of them as tiny spheres at each point of space, or even tiny toruses (doughnut shapes). But things become more difficult when we go to more dimensions. Furthermore, we would like to be able to visualize these higher dimensions. Fortunately, in 1984 Philip Candelas of the University of Texas at Austin, Andrew Strominger of the University of California at Santa Barbara, and Edward Witten showed that the six extra dimensions of superstring theory could be represented by what are called Calabi-Yau shapes. Such shapes (or spaces, as we frequently refer to them) were devised by Eugene Calabi of the University of Pennsylvania and Shing-tung Yau of Harvard University. Their original work had nothing to do with strings. An example of a Calabi-Yau shape is shown in figure 65. Again we have to think of these figures as at each point in space.

A PLETHORA OF THEORIES

Supersymmetry gave string theory a boost. Incorporating it into string theory solved many problems, but it also caused a dilemma. It turned out that supersymmetry could be incorporated into the theory in five different ways, and each led to a different superstring theory. As a result, we now have five different superstring theories. The details of these theories are not of importance to us here, but, for completeness, I will give a brief description of each of them.[12]

Type	Description
I	Superstring theory between forces and matter that contains both open and closed strings. No tachyon. Based on group SO(32).
IIA	Superstring theory between forces and matter with only closed strings (loops). No tachyon and massless fermions spin both ways.
IIB	Superstring theory between forces and matter with closed strings only. No tachyon and massless fermions spin in only one way.

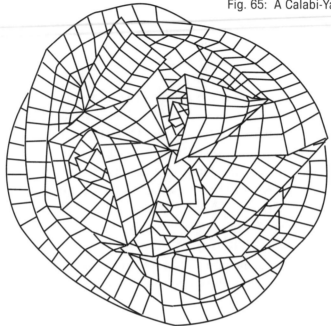

Fig. 65: A Calabi-Yau shape.

HO   Superstring theory between forces and matter with closed strings only. No tachyon. Heterotic, meaning that right and left moving strings differ. Based on the group S(32).

HE   Superstring theory between forces and matter with closed strings only. No tachyon. Heterotic and based on the group $E_8 \times E_8$.

The existence of five different theories is obviously a problem. After all, we are looking for a single theory, which we hope will be a theory of everything (TOE). So what is wrong? Is one of the above theories correct, and the other four incorrect? Or are they just different versions of the same theory? We will consider these questions in the next chapter.

TESTING THE THEORY

One of the time-honored traditions of science is experimental verification of predictions that are made in a theory. If the theory doesn't predict something that can be measured and verified, it isn't much of a theory. At least this is what was assumed in the past. The problem with string theory is that the strings are so short (10^{-33} centimeter) that they are far beyond our technical ability to measure. At the present time, we can probe to a billionth of a billionth of a meter. This probing is done with accelerators, and it's easy to show that we would need an accelerator the size of the universe to probe down to the size of a string. What does this leave us with? Does it mean we're never going to be able to test string theory? No, but it does mean that we're going to have to test it indirectly, and, fortunately, there are several possibilities for accomplishing this.

One approach is to look at the consequences of the extra dimensions curled up in a Calabi-Yau shape. Philip Candelas, Gary Horowitz, Andrew Strominger, and Ed Witten have shown that there are predictions that we may be able to obtain from the various Calabi-Yau patterns that are possible. No one has been able to follow through on this yet; nevertheless, it is a possible way of testing the theory and may be used in the future.

Another approach is to consider the number of families of particles. We know there are three, but we have no idea why. Witten and others have shown that one family is associated with each of the holes in the Calabi-Yau space. If the curled-up Calabi-Yau space has three holes, it would mean there should be three families. So far, however, physicists have not been able to determine how many holes there are.

One of the best tests would be the detection of the superpartners. They are believed to be just beyond our present accelerators. Their detection would be a great help, but it would not necessarily prove that string theory is correct. On the other hand, if we don't detect them with the next generation of accelerators, it doesn't necessarily rule out string theory. They might still be beyond their range.

There are also several other possible tests associated with the early universe and cosmology. Also, there is a test associated with the predicted fractional charges of the theory. Observation of them would be helpful. In conclusion, experimental verification is no doubt going to be difficult, but, as we have seen, it is not impossible.

Chapter 12

Beyond Superstrings

M-Theory

In the early 1990s we had five apparently independent string theories. They shared many features but were distinctly different in the way they incorporated supersymmetry. The available vibrational patterns in each theory, for example, were different. We were, of course, searching for a single theory that would encompass all four forces of nature and explain the elementary particles, so this was obviously a problem. We didn't want our "ultimate" theory to consist of five theories; it wouldn't make any sense. The only way around this was either four of the theories are incorrect and one is correct, or all five are somehow equivalent. If neither of these were true, we were obviously in trouble, and string theory was the wrong approach altogether. Part of the difficulty, it was soon realized, was the way we were tackling the problem. The equations were difficult to solve, and the only thing we could do was use approximations, and these approximations were a problem. Finally, however, in 1995 a breakthrough came that shed new light on the problem. To understand this breakthrough, we have to look briefly at what is called *perturbation theory*.

THE STANDARD APPROACH: PERTURBATION THEORY

By the mid-1990s scientists were finally beginning to realize that the approach they were using was flawed. It wasn't an unreasonable approach; indeed, it was the natural one to take at the time. It is referred to as perturbation theory and is used in all major quantum theories, such as quantum electrodynamics, quantum chromodynamics, and electroweak theory. To illustrate it, we will consider quantum electrodynamics.[1] Suppose an interaction has just taken place—say, the collision of an electron and a positron as shown in figure 66.

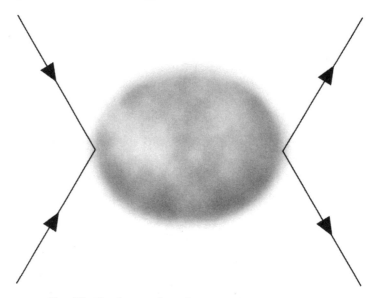

Fig. 66: The interaction of two particles. We show it
as a cloud since we don't know what happens.

We see the particles that go into the interaction and those that come out, but we're not quite sure what happens in between, so we show it as a cloud.

One of the key elements in such interactions is what is called the *coupling constant*.[2] It is a number that tells us how strong the interaction is. Each of the forces of nature has a coupling constant; the one for the electromagnetic field is 1/137. We therefore expect that superstring theory would also have a coupling constant; in fact, each of the five superstring theories discussed above should have such a constant.

This coupling constant is critical in perturbation theory. Let's begin with an overview of perturbation theory. The idea is to get a ballpark approximation, then narrow in on a more exact result. We refer to the first approximation as first-order perturbation theory, the second approximation as second-order perturbation theory, and so on. When we have what we want, we add all the contributions together, and this gives us (hopefully) a highly accurate result. It's something like what happens when your refrigerator breaks down. You ask the repairman how much it will cost. He gives you an estimate of $200. Later, when he's into the job, he says it's going to cost a little more, say $240, and when you finally get the bill, it says $244.50.

As it turns out, perturbation theory works only if the coupling constant is less than one. This is why knowledge of this constant is so important. For quantum electrodynamics this is obviously not a problem, since it is only 1/137. Let's look at how we would apply the procedure to the interaction described above, namely, the collision of an electron and a positron. In quantum electrodynamics we use what are called Feynman diagrams, named for the brilliant physicist Richard Feynman who first used them.[3]

The first-order Feynman diagram for the above process looks as follows.

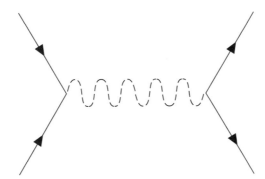

Fig. 67: Feynman representation of the interaction.

Using this diagram, we can calculate the first-order effect, and we arrive at a number. We then look at second-order effects; there are several diagrams in this case. One of them looks as follows.

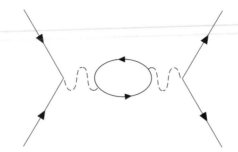

Fig. 68: A second-order Feynman diagram.

Again, from this diagram we get another contribution to our overall effect, and we add it to the first-order effect. In theory, we can continue this way to third-order effects, although in practice it is rarely used.

When we use the perturbation approach in superstring theory, we replace the above diagrams with string diagrams, which look as follows.

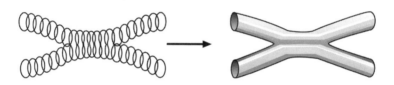

Fig. 69: String representation of a Feynman diagram.
The more complete world sheet is shown on the right.

We then proceed in the same way as we did previously for first-order effects. For higher-order results we have more complicated diagrams such as[4]

Fig. 70: A higher-order Feynman diagram.

And again we calculate the contribution from them. In theory, everything sounds straightforward, but in practice there are serious problems. First of all, the calculations are extremely difficult, and, even worse, we don't know the string theory coupling constant. Moreover, the procedure only works when this constant is less than one. If it is greater, our calculations would have been in vain, since they would be meaningless.

String theory seemed to have hit an impasse. Something had to be done. And in 1995 there was a breakthrough.

THE SECOND SUPERSTRING REVOLUTION

In March 1995 the annual superstring conference was held at the University of Southern California.[5] There was considerable interest, since several new developments had occurred during the year, but no one expected the bombshell that came. Everyone was looking forward to the lecture of the acknowledged leader of superstring theory, Ed Witten. But even with all the anticipation, what he had to say hit like a thunderbolt. Several people had suggested that the perturbation theory approach was a dead-end street and that something more was needed. But no one had suggested how it could be bypassed. Witten showed that it could.[6] The key, he said, was dual relationships, or *dualities*. He stated that, through the use of dualities, he could show that the five superstring theories were closely related. In effect, they were just five different ways of expressing the same physics.

As we saw earlier, the success of the perturbation approach depends on the size of the coupling constant. It has to be less than one; we refer to this as *weak coupling*. If it is greater than one, we have *strong coupling*. And the problem with superstring theory was that we didn't know what the constant was. Because of this, it seemed best to examine the various theories for different values of the coupling constant. For low values we could use perturbation theory, but there was a problem for high coupling constants. Nevertheless, as Witten pointed out, there was a way of determining some of the masses and charges of the theory when the coupling was strong. It was discovered by Manoj Prasad, E. Bogomoln'yi, and Charles Sommerfeld. The states they discovered are now referred to as PBS states, and with them we had at least a slight handle on the strong coupling case.

Fig. 71: Ed Witten.

Witten began with Type I superstring theory. His idea was to explore
the theory for a variety of coupling constants, ranging from very weak to
strong. To his surprise, he found that Type I theory with a coupling constant greater than one agreed exactly with Heterotic-O string theory with
a coupling constant less than one. In other words, they were exactly the
same for this case. This meant that we could make calculations in Heterotic-O theory with a weak constant (where perturbation theory was
valid), and we would know the answers for similar calculations in Type I
string theory with a strong coupling constant. The physics was identical
in the two cases.

Witten then showed that this "duality," as it was called, was also valid for Type IIB string theory. In this case, however, things were a little different. The duality was between weak and strong coupling in the same theory. In other words, results from weak coupling in Type IIB theory were the same as for strong coupling. This meant the theory was *self-dual*. We now refer to this and the previous dualities as S-dualities.

Everyone at the conference was stunned by Witten's announcement. It meant that the five theories were connected, or at least some of them were connected. But the biggest bombshell was yet to come.

LINK TO SUPERGRAVITY

In an earlier section, we discussed a theory called supergravity. It was an eleven-dimensional theory that had been around for several years. Basically, as in superstring theory, supersymmetry had been added to general relativity in the hopes that it might lead to a theory that would give us some new insights. It referred to point particles, so it had nothing to do with string theory. In the end it was not considered a success, but there was still considerable interest in the theory.

Witten showed that there was a link between the five superstring theories and supergravity.[7] He showed that if we start with Type IIA superstring theory, and increase the coupling constant until it is larger than one, the physics goes over to that in a low-energy approximation to eleven-dimensional supergravity. This was particularly surprising. How could a ten-dimensional theory link with an eleven-dimensional theory? It didn't seem to make sense.

Witten showed, however, that it did make sense. He pointed out that there was actually an unseen dimension in superstring theory that hadn't been noticed. As he increased the coupling constant in Heterotic-E theory, the one-dimensional string loops within the theory began to look like tubes. In other words, they became two-dimensional membranes. No one had noticed this before because all the calculations had been done using perturbation theory and a small coupling constant. More explicitly, he found that the strongly coupled Heterotic-E string theory had an eleven-dimensional description that was equivalent to eleven-dimensional supergravity. With supergravity, there were now six theories that were linked.

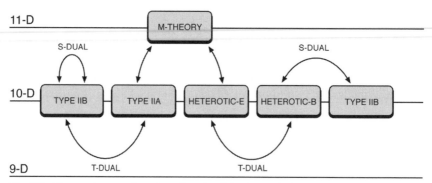

Fig. 72: The various string theories (and M-theory) shown in ten and eleven dimension. The S and T dualities are shown.

SMALL-LARGE DUALITIES

These developments generated a lot of excitement, but the picture was not yet complete. In particular, all the theories were not linked. But this was not the end of the story; as it turned out, there were more dualities. To see how they arise, consider the compactification, or "curling up," of one of the dimensions in superstring theory. Assume that it is wound into a small-diameter cylinder. It's easy to see that strings can move on this cylinder in several ways: they can wrap themselves around it, or they can move across it in "unwrapped" modes as shown in figure 73. This tells us that the energy of the strings comes from two sources: vibrational motion and winding energy. This leads to what are now called small-large dualities.[8] As was shown by Witten and others, for a radius R in our cylindrical universe, there is a smaller circular radius for which the winding energies of the strings equal the vibrational energies of a large-radius universe. In short, there was a duality between R and $1/R$. It is referred to as a T-duality. Another way of stating this is to say that exchanging vibrational modes of the string with winding modes exchanges a large distance scale with a small distance scale.

The string theories associated with this duality are Type IIA and Type IIB. This means if we make calculations in Type IIA string theory for circular radius R (large scales), we have the answers for circular radius $1/R$ (small scales). There is a similar duality between Heterotic-E string theory and Heterotic-O theory. In summary, then, the physics for a Type

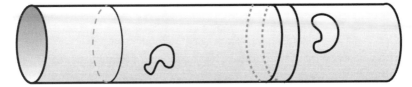

Fig. 73: Strings on a cylinder.

IIA string in a universe of radius R is identical to the physics of a Type IIB string in a universe with a circular radius of $1/R$.

This completes the linking of the five superstring theories. They are, in effect, all connected by a web of dualities.

M-THEORY

Witten called the new theory *M-theory*.[9] No one is certain exactly what the M stands for. Various candidates are "Mother of all theories," "Mystery theory," "Membrane theory," and "Matrix theory." All five of the string theories, along with eleven-dimensional supergravity, are linked by M-theory. We can pass from one theory to the other via the various dualities described above. This is shown in figure 74. Note that if we start in one of the appendages, say, Heterotic-O or Type IIB, and turn up the coupling constant, we move toward the center of the figure and in the process the one-dimensional strings of the outer regions become two-dimensional membranes. All five theories, in fact, have two-dimensional membranes. This leads us to ask: which are the more fundamental—strings or membranes? As it turns out, in the weak coupling case, the membranes are extremely massive, much more massive than the Planck mass, and this is a problem when it comes to explaining the particles. There are, however, certain cases where this does not happen.

M-theory has also made an important contribution to cosmology. As we saw earlier, the three nongravitational forces were all equal in the very early universe. It was suspected that at sufficiently early times (and high energies), gravity was also equal to the other three, but calculations showed that it was not *exactly* equal. With M-theory, however, all four of the forces came together and were of equal value.

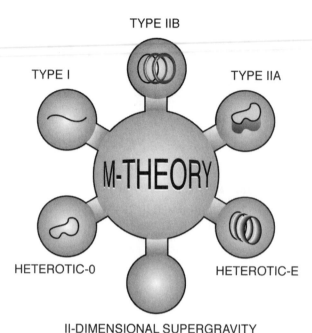

TYPE IIB

TYPE I   TYPE IIA

M-THEORY

HETEROTIC-0   HETEROTIC-E

II-DIMENSIONAL SUPERGRAVITY

Fig. 74: M-theory and its relationship to the various superstring theories.

If there are two-dimensional membranes, we immediately wonder about higher-order ones. Is it possible there are three- and four-dimensional membranes? Indeed, it is, and according to the theory, they exist. In fact, membranes up to nine dimensions are possible.

P-BRANES AND THE BRANE WORLD

Paul Townsend of Cambridge University, who did much of the early work on membranes, named them *branes,* and since they can have any dimension up to nine, he refers to the various higher-dimensional ones as p-branes.[10] Thus we have 1-branes, which are strings, 2-branes, which are membranes or surfaces, and so on.

What exactly is a brane? Simply described, it is a spacetime object that is a solution to Einstein's equation in the low-energy limit of superstring theory. Taken together, the branes give us a "brane world." In this brane world the strong, electromagnetic, and weak forces are trapped on a three-dimensional brane. We can view it as an exceedingly thin surface. Gravity is not trapped by this brane; it is free to roam the region around the brane and can move through this volume. Indeed, it is assumed that the gravitational force is so weak because it is diluted by the additional dimensions.

Scientists are just beginning to understand branes, the brane world, and M-theory. So far they have barely touched the surface, and there is much to be learned.

ANOTHER BLACK HOLE CONNECTION

Are black holes connected with superstrings? Or, more generally: are they connected with M-theory and branes?[11] At first, this might seem like a ridiculous question. After all, black holes are huge compared to strings. Stellar-collapse black holes are a few miles across, and strings are billions of billions of times smaller than the nucleus. But, interestingly, in recent years a connection has been found.

Let's look at things from a simple point of view first. We know that we can describe a black hole in terms of only three properties: mass, charge, and spin. We saw earlier that this gives us four distinct types of black holes. Looking at elementary particles, we see that they also are identified by three properties, namely, mass, charge, and spin. These are obviously the same three that describe black holes. And, according to string theory, elementary particles are just various states of vibration of strings. So there is a link between black holes and strings.

We also know that in addition to stellar-collapse black holes there are primordial black holes, which were presumably produced in the big bang explosion. And there is no lower limit to the mass of these black holes. In fact, there is no reason why they can't be as small as elementary particles or strings. Furthermore, as we saw earlier, black holes evaporate according to Hawking, and this evaporation eventually leads to elementary particle–sized black holes. So we do, indeed, appear to have black holes as small as elementary particles.

One of the key predictions that led Hawking to his discovery of the evaporation of black holes was one that came in the early 1970s. Jacob Bekenstein, who was then at Princeton University, predicted that black holes contained entropy, and a measure of this entropy was the surface area of the black hole. This led to a new science referred to as black hole thermodynamics. But an enigma soon developed. Hawking said that black holes evaporate by giving off radiation, and as they did, they got hotter and smaller. Eventually, they exploded and disappeared. If this is true, we have to ask: what happens to their entropy? It's easy to show that there is a tremendous amount of entropy tied up in a black hole. No one could explain this adequately.

A breakthrough came in 1996 when Andrew Strominger and Cumrun Vafa published (more exactly—posted on the Internet) a paper titled "Microscopic Origin of Bekenstein-Hawking Entropy."[12] They directed their result at what are called extreme black holes of electrical charge (extreme black holes have the maximum amount of charge that they can hold). They showed that they could "build" extreme black holes of electrical charge using branes. What they did was bring together branes to produce a black hole. They then counted the number of rearrangements of the constituents of their black hole to obtain its entropy. To their surprise, they found that it agreed exactly with the entropy predicted by Bekenstein and Hawking. Thus, using string theory, they were able to get the same result for black holes as was obtained through conventional black hole physics. This amazed physicists and was considered to be a significant result.

HOLOGRAMS AND BRANES

We introduced the idea of holograms in the last chapter, and, as we will see, they might also play an important role in relation to branes.[13] A hologram, as noted earlier, encodes the information from a region of space onto a surface, so in the case of branes, it codes this information from a p-brane to a (p-1) brane. This means that in our four-dimensional world there is a one-to-one correspondence between the states of our world and a five-dimensional world, if the hologram hypothesis is true.

THE FINAL CONSENSUS

Dramatic advances have been made in recent years in superstring theory. But as we have seen, dramatic advances have also been made using a different approach, namely, loop quantum gravity. Both theories give us a union of the four basic forces of nature and explain elementary particles. But both theories are also relatively unexplored and at a primitive stage. We still do not know many of the implications of M-theory or of loop quantum gravity.

In a sense, we have the same problem we had with superstring theory

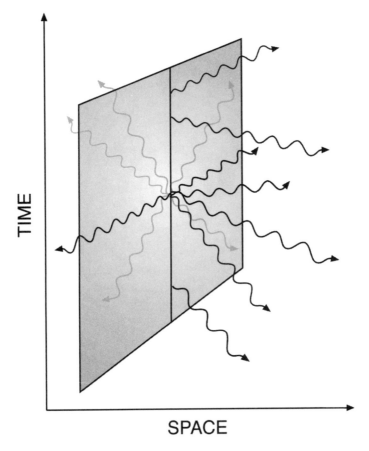

Fig. 75: Matter and nongravitational forces are confined to the brane (flat surface). Gravity can spread out to extra dimensions.

a few years ago with the five different theories. Scientists eventually showed that they are just different windows on one central theory. Now the question is: which is the correct theory, M-theory or loop quantum gravity? Is one wrong and one right, or is it possible that they are just two different ways of looking at the problem? Are they equivalent in the way Schrödinger's and Heisenberg's formulations of quantum mechanics are? Both theories have the same connections to black holes and holograms.

Lee Smolin, who has strong ties to quantum loop gravity, believes it is more likely to be the true theory. He says that string theory has a weakness in that it is not "background independent." This idea is central to general relativity, which specifies that there is no fixed background, and since loop quantum gravity is based on general relativity, it is background independent. On the other hand, many people in the string camp are skeptical of loop quantum gravity and are sure that M-theory will turn out to be the long-awaited theory of everything.

Epilogue

This concludes our look at Einstein's remarkable discoveries and how they have shaped modern science. His insights have, without a doubt, had a tremendous influence on modern physics, and on astronomy, cosmology, and even chemistry, in that they have helped shape these sciences and have led to many new branches of physics such as black hole physics and gravitational wave physics. It is hard not to feel a sense of awe as you look back over Einstein's numerous contributions. His greatest discoveries were, of course, his special and general theories of relativity. They came early in his life. Both were radical at the time, so radical that many people refused to believe them. Even as late as 1922, when the Nobel Prize committee finally realized, after years of procrastination, that they had to award the Nobel Prize to Einstein, they still refused to give it to him for his theory of relativity. They gave it to him instead for his early contributions to quantum theory and the explanation of the photoelectric effect.

Einstein, however, did not stop with his two relativity theories. Immediately after publishing his general theory of relativity, he applied it

Fig. 76: Einstein.

to the universe and gave us one of the first cosmologies. It was at this time
that he introduced his famous "cosmological constant," and even though
he eventually discarded it, others kept it, and it remained controversial for

years. Then, with the recent discovery that the expansion of the universe may be accelerating, it took center stage, and it may again end up playing a central role in cosmology.

Not only were Einstein's relativity theories radical, so, too, were some of the objects that were predicted by them. Two of the most fascinating and bizarre were black holes and wormholes in space. Einstein didn't like the idea of black holes, and he tried to show that they couldn't exist. But J. Robert Oppenheimer proved that they were real; he demonstrated that they would form when massive stars exhausted their fuel and collapsed. Interestingly, wormholes have generated even more interest than black holes lately. Scientists have shown that time travel may be possible using them. As every science fiction reader knows, travel to the future and past has long been a staple of that genre, but most scientists thought it was only a pipe dream. We now know, however, that it may one day be possible.

Another of Einstein's predictions that we explored was gravitational waves. They are waves that are produced when black holes, or neutron stars, collide; supernova explosions also generate them. Over the years scientists have searched for these waves without success, but recently several gigantic gravitational wave detectors have been built, and within a few years even larger ones will be assembled in space. Scientists are confident that gravitational waves will soon be found.

Another of Einstein's predictions—cosmic lenses—has become an important tool in astronomy. According to this prediction, a distant object —a galaxy or black hole—can magnify an object behind it. Several of these lenses have now been found, and the lensing technique, as it is called, is now being used to search for extrasolar planets.

Einstein's genius is abundantly clear in all of his discoveries, but we also saw that he could be stubborn. Despite the tremendous successes of quantum theory, he refused to fully accept it. It wasn't its predictions that bothered him; it was the theory's weird philosophical implications that left him cold. He never accepted them, but recent experiments have shown that they are, indeed, valid.

Einstein was not happy about all his contributions. As we noted, his famous formula $E = mc^2$ became the basis of the atomic and hydrogen bombs, and as a strong pacifist he was horrified. Although he never played a direct role in the building of the bomb, he felt guilty about it for the rest of his life.

Einstein was also indirectly responsible for several things that are not usually associated with his name. Lasers and masers, for example, now play a major role in our everyday affairs. The original paper showing that population inversions were possible was written by Einstein, and they are what make lasers and masers possible. He also showed that superfluidity could be attained at extremely low temperatures, and he predicted what is now called the Bose-Einstein condensate. Scientists have recently produced it in the lab.

People have said that the last thirty years of Einstein's life were wasted in a futile attempt to find a unified field theory. And, indeed, he searched relentlessly for such a theory, but he never found one. To say that the years were wasted, however, does not do them justice; in reality, Einstein was, in fact, ahead of his time. Many people felt that he was chasing a rainbow, but the search for such a theory, or more generally, a theory of everything (TOE), has now become a major endeavor within the physics community. Hundreds of physicists around the world are now following in his footsteps, searching for the ultimate theory, and, within the last few years, have made considerable progress. Two approaches have now come to the forefront: loop quantum gravity and superstring theory. Both are viable candidates for a TOE, but problems still need to be overcome. Superstring theory has actually been superceded now by M-theory, and it has generated a lot of enthusiasm.

Will Einstein's dream ever be achieved? Will a final theory be found? Einstein certainly was optimistic. Even when he was on his deathbed, he had the nurse bring him his notepad and pencil, and it was soon filled with equations. He confided to his friend Otto Nathan the day before he died that he was sure he was close to success.

Einstein's dream of a theory of everything may, indeed, one day be achieved. If you think about it, however, finding it could prove disheartening. A true and final TOE would be a disappointment for the generations of physicists that came after it. There would be nothing left to discover. It seems more likely that there will always be problems to solve and mysteries to enchant us, with theories getting closer and closer to the ultimate truth, but never quite reaching it. As the past has shown, the future is hard to predict, and there will always be surprises. I believe that this will continue to be the case.

Notes

INTRODUCTION

1. Albrecht Fölsing, *Albert Einstein: A Biography* (New York: Viking, 1997), p. 176.

2. Albert Einstein, *Relativity: The Special and General Theory,* trans. Robert W. Lawson (New York: Crown, 1961).

3. Barry Parker, *Cosmic Time Travel: A Scientific Odyssey* (New York: Plenum, 1991), p. 112.

4. Ibid., p. 125.

5. Nathan Cohen, *Gravity's Lens: Views of the New Cosmology* (New York: Wiley, 1988), p. 145.

6. Fölsing, *Albert Einstein: A Biography*, p. 722.

7. Ibid., p. 576.

8. Brian Greene, *The Elegant Universe: Superstrings, Hidden Dimensions, and the Quest for the Ultimate Theory* (New York: Norton, 1999), p. 135.

CHAPTER 1. TWISTS IN THE FABRIC OF SPACE

1. Denis Brian, *Einstein: A Life* (New York: Wiley, 1996), p. 71.
2. Ibid., p. 79.
3. Don Howard and John Stachel, *Einstein and the History of General Relativity* (Boston: Birkhauser, 1989), p. 164.
4. Ibid., p. 49.
5. Albrecht Fölsing, *Albert Einstein: A Biography* (New York: Viking, 1997), p. 245.
6. Ibid., p. 314.
7. Howard and Stachel, *Einstein and the History of General Relativity*, p. 127.
8. Ibid., p. 186.
9. Fölsing, *Albert Einstein: A Biography*, p. 364.
10. Ibid., p. 373.
11. Ibid., p. 374.
12. John Stachel, *Einstein from B to Z* (Boston: Birkhauser, 2002), p. 353.
13. Howard and Stachel, *Einstein and the History of General Relativity*, p. 213.
14. Paper is republished in Albert Einstein et al., *The Principle of Relativity* (New York: Dover, 1923).

CHAPTER 2. EXPANDING TO SPACE: HISTORY OF THE COSMOLOGICAL CONSTANT

1. Albert Einstein et. al., *The Principle of Relativity* (New York: Dover, 1923), p. 179.
2. Ibid., p. 184.
3. Ann Hentschel, trans., Robert Schumann et al., eds., *The Berlin Years, Correspondence, 1914–1918*, vol. 8 of *The Collected Papers of Albert Einstein*, English translation (Princeton: Princeton University Press, 1987–1998), p. 309.
4. Edwin Hubble, "A Relation between Distance and Radial Velocity among Extra-Galactic Nebulae," *Proceedings of the National Academy of Science* 15, no. 3 (1929): 168.
5. Edwin Hubble, *Realm of the Nebulae* (Yale: Yale University Press, 1936; reprint, New York: Dover, 1958).
6. Albrecht Fölsing, *Albert Einstein: A Biography* (New York: Viking, 1997), p. 390.

7. There is considerable information on the satellite COBE in Barry Parker, *The Vindication of the Big Bang: Breakthroughs and Barriers* (New York: Plenum, 1993).

8. Barry Parker, *Invisible Matter and the Fate of the Universe* (New York: Plenum, 1989), p. 95.

9. Jeremiah Ostriker and Paul Steinhardt, "The Quintessential Universe," *Scientific American* (January 2001): 47.

10. João Magueijo, "Plan B for the Cosmos," *Scientific American* (January 2001): 58.

11. Mordehai Milgrom, "Does Dark Matter Really Exist?" *Scientific American* (August 2002): 42.

CHAPTER 3. BLACK HOLES, WORMHOLES, AND OTHER DEMONS

1. Don Howard and John Stachel, *Einstein and the History of General Relativity* (Boston: Birkhauser, 1989), p. 216.

2. A. S. Eddington, *The Internal Constitution of the Stars* (New York: Dover, 1959).

3. Kip Thorne, *Black Holes and Time Warps: Einstein's Outrageous Legacy* (New York: Norton, 1994), p. 135.

4. They were first called "black holes" by John Wheeler.

5. Thorne, *Black Holes and Time Warps*, p. 122.

6. Further details of the life of Oppenheimer can be found in Peter Goodchild, *J. Robert Oppenheimer: Shatterer of Worlds* (Boston: Houghton Mifflin, 1981).

7. The first such radio source was discovered by Allan Sandage of Hale Observatory using the two-hundred-inch Hale reflector in 1960. Another one was found in 1962, and astronomers were able to narrow in on it and observe its optical counterpart. To their surprise, it looked like an ordinary star. In 1963 Maarten Schmidt of Caltech determined that the spectral lines of the object had undergone a tremendous redshift. This placed the object near the edge of our observable universe.

8. B. K. Harrison et al., *Gravitation Theory and Gravitational Collapse* (Chicago: University of Chicago Press, 1963).

9. Actually, because of the curvature of space caused by the black hole, stars directly behind the black hole can be seen. Their images are forced together in the region near the perimeter of the black hole.

10. The discovery was actually made by one of Newman's students. The student pointed out that the transformation from the Schwarzschild black hole to the Kerr black hole (which was known) could be applied to the Reissner-Nördstrom black hole, and it would give a solution for the spinning, charged black hole.

11. It has, in fact, been shown that if you could capture a small black hole and put it in orbit around Earth, you could use it to generate electricity, which could be beamed to Earth. All you would have to do is project particles into the ergosphere in just the right way.

12. Further details on Hawking's life can be found in John Boslough, *Stephen Hawking's Universe* (New York: Avon, 1985).

13. A more detailed description of the concept of entropy can be found in Thorne, *Black Holes and Time Warps*, p. 422.

14. As we will discuss later in the book, the link between general relativity and quantum theory is one of the foremost problems in physics at the present time.

15. Barry Parker, *Cosmic Time Travel: A Scientific Odyssey* (New York: Plenum, 1991), p. 198.

16. Many books contain detailed discussions of Cyg X-1. One of the best is William Kaufmann, *The Cosmic Frontiers of General Relativity* (Boston: Little, Brown 1977), p. 228.

CHAPTER 4. THE MYSTERY OF TIME AND TIME TRAVEL

1. Paul Davies, "That Mysterious Flow," *Scientific American* (September 2002): 41.

2. Ibid., p. 40.

3. Barry Parker, *Cosmic Time Travel: A Scientific Odyssey* (New York: Plenum, 1991), p. 93.

4. Carl Sagan, *Contact* (New York: Pocket Books, 1985). The book was also made into a movie starring Jodie Foster.

5. The paper published by Morris and Thorne is "Wormholes in Spacetime and Their Use for Interstellar Travel," *American Journal of Physics* 56 (May 1988): 395.

6. A more detailed discussion of the Casimir effect is found in Jim Al-Khalili, *Black Holes, Wormholes, and Time Machines* (Bristol: Institute of Physics, 1999).

7. Parker, *Cosmic Time Travel*, p. 226.

8. Al-Khalili, *Black Holes, Wormholes, and Time Machines*, p. 220.

9. Parker, *Cosmic Time Travel*, p. 230.

10. There is a good discussion of causality in William Kaufmann, *The Cosmic Frontiers of General Relativity* (Boston: Little, Brown, 1977), p. 240.

11. The idea of parallel universes is discussed in John Gribbin, *Timewarps* (New York: Delacortes, 1979), p. 111.

12. Strings are discussed in chapters 11 and 12. An excellent reference on string theory is Brian Greene, *The Elegant Universe: Superstrings, Hidden Dimensions, and the Quest for the Ultimate Theory* (New York: Norton, 1999).

13. Lawrence Ford and Thomas Roman, "Negative Energy, Wormholes, and Warp Drive," *Scientific American* (January 2000): 46.

14. Stephen Hawking, *The Universe in a Nutshell* (New York: Bantam, 2001), p. 64.

15. One of the best references on supercivilizations is Thomas McDonough, *The Search for Extraterrestrial Intelligence: Listening for Life in the Cosmos* (New York: Wiley, 1987).

CHAPTER 5. RIPPLES IN THE CURVATURE OF SPACE

1. Albert Einstein, "On Gravitational Waves," *SBI* (*Proceedings of the Prussian Academy*) (1918): 154.

2. Albert Einstein and Nathan Rosen, "On Gravitational Waves," *Journal of the Franklin Institute* 223 (1937): 43.

3. There is considerable background material on Weber in Marcia Bartusiak, *Einstein's Unfinished Symphony: Listening to the Sounds of Space-Time* (Washington, DC: Joseph Henry Press, 2000), p. 90.

4. It was believed at the time (and now) that the center of our galaxy was a very violent place and therefore a likely source of gravitational waves. We now believe, in fact, there may be a massive black hole there.

5. Considerable background on these detectors and their inventors can be found in Kip Thorne, *Black Holes and Time Warps: Einstein's Outrageous Legacy* (New York: Norton, 1994).

6. A detailed discussion of the gravitational waves emitted in the collision of two black holes is given in Thorne, *Black Holes and Time Warps*.

7. Barry Parker, *Cosmic Time Travel: A Scientific Odyssey* (New York: Plenum, 1991), p. 187.

8. Bartusiak, *Einstein's Unfinished Symphony*, p. 205.

9. Ibid., p. 74.

10. Thorne, *Black Holes and Time Warps*, p. 383.

11. Ibid., p. 370.

12. W. Gibbs, "Ripples in Spacetime," *Scientific American* (April 2002): 63.
13. Ibid.
14. Ibid.

CHAPTER 6. GRAVITY'S COSMIC LENSES

1. Hans Ohanian and Remo Ruffini, *Gravitation and Spacetime* (New York: Norton, 1994), p. 203.

2. Albert Einstein, "Lens-like Action of a Star by Deviation of Light in the Gravitational Field," *Science* 84 (1936): 506.

3. Nathan Cohen, *Gravity's Lens: Views of the New Cosmology* (New York: Wiley, 1988), p. 158.

4. Joachim Wambsganss, "Gravity's Kaleidoscope," *Scientific American* (November 2001): 68.

5. D. Walsh, R. F. Carswell, and R. J. Weyman, "0957 + 561 A, B—Twin Quasistellar Objects or Gravitational Lens," *Nature* 279 (1979): 381.

6. Barry Parker, *Invisible Matter and the Fate of the Universe* (New York: Plenum, 1989), p. 195.

7. Wambsganss, "Gravity's Kaleidoscope," p. 65.

8. Ibid., p. 68.

9. Ibid., p. 69.

10. Ibid., p. 68.

11. Ibid., p. 70.

12. Ibid.

13. Ibid.

14. Ibid., p. 71.

15. Ibid.

16. Cohen, *Gravity's Lens*, p. 211.

CHAPTER 7. EINSTEIN'S QUANTUM LEGACY

1. Alice Calaprice, *The Expanded Quotable Einstein* (Princeton: Princeton University Press, 2000), p. 260.

2. Ibid., p. 245.

3. Albert Einstein, Boris Podolsky, and Nathan Rosen, "Can Quantum Mechanical Description of Physical Reality Be Considered Complete?" *Physical Review* 47 (1935): 797.

4. Einstein was awarded the Nobel Prize in 1922, but it was for the year 1921.

5. See Barry Parker, *Quantum Legacy: The Discovery That Changed Our Universe* (Amherst, NY: Prometheus Books, 2002) for a more detailed discussion.

6. Ibid., p. 121.

7. Ibid., p. 130.

8. Einstein, Podolsky, and Rosen, "Can Quantum Mechanical Description of Physical Reality Be Considered Complete?"

9. Ibid.

10. Albrecht Fölsing, *Albert Einstein: A Biography* (New York: Viking, 1997), p. 697.

11. Ibid., p. 698.

12. Amir Aczel, *Entanglement: The Greatest Mystery in Physics* (New York: Four Walls Eight Windows, 2002), p. 122.

13. John von Neumann, *The Mathematical Foundations of Quantum Mechanics,* trans. Robert T. Beyer (Princeton: Princeton University Press, 1955).

14. Aczel, *Entanglement,* is an excellent source of information on John Bell.

15. Ibid., p. 149.

16. Ibid., p. 177.

17. Anton Zeilinger, "Quantum Teleportation," *Scientific American* (April 2000): 50.

18. Ibid.

19. Barry Parker, *Cosmic Time Travel: A Scientific Odyssey* (New York: Plenum, 1991), p. 276.

20. Denis Brian, *Einstein: A Life* (New York: Wiley, 1996), p. 164

CHAPTER 8. SUPERBOMBS

1. Two books have recently been written with the title "$E = mc^2$."

2. Albert Einstein et al., *The Principle of Relativity* (New York: Dover, 1923), p. 69.

3. Barry Parker, *Quantum Legacy: The Discovery That Changed Our Universe* (Amherst, NY: Prometheus Books, 2002), p. 29.

4. Albrecht Fölsing, *Albert Einstein: A Biography* (New York: Viking, 1997), p. 707.

5. Ibid., p. 710.

6. Parker, *Quantum Legacy*, p. 216.

7. David Bodanis, *E = mc²: A Biography of the World's Most Famous Equation* (New York: Berkeley Books, 2000), p. 134.

8. Richard Rhodes, *The Making of the Atomic Bomb* (New York: Simon and Schuster, 1986).

9. Bodanis, *E = mc²*, p. 156.

10. Fölsing, *Albert Einstein: A Biography*, p. 715.

11. Ibid., p. 716.

12. Ibid., p. 720.

13. Richard Rhodes, *Dark Sun: The Making of the Hydrogen Bomb* (New York: Simon and Schuster, 1996), p. 252.

14. Ibid., p. 207.

15. Ibid., p. 208

16. Fölsing, *Albert Einstein: A Biography*, p. 726.

CHAPTER 9. OTHER EINSTEIN INSIGHTS

1. John Stachel, *Einstein from B to Z* (Boston: Birkhauser, 2002), p. 520.

2. Ibid., p. 521

3. Ibid., p. 524.

4. Albert Einstein, "Quantum Theory of the Single-Atom Ideal Gas," *SB* (*Proceedings of the Prussian Academy*) (1924): 261.

5. Eric Cornell and Carl Wieman, "The Bose-Einstein Condensate," *Scientific American* (March 1998): 40.

6. Ibid.

7. Gene Dannen, "The Einstein-Szilard Refrigerator," *Scientific American* (January 1997): 90.

8. Ibid.

9. Ibid.

10. Barry Parker, *Quantum Legacy: The Discovery That Changed Our Universe* (Amherst, NY: Prometheus Books, 2002), p. 159.

11. Ibid., p. 168.

12. John Stachel, *Einstein's Miraculous Year: Five Papers That Changed the Face of Physics* (Princeton: Princeton University Press, 1998), p. 161.

13. Barry Parker, *Einstein: The Passions of a Scientist* (Amhert, NY: Prometheus Books, 2003), p. 146.

14. Ibid., p. 148.

CHAPTER 10. DREAMS OF A UNIFIED THEORY

1. Albrecht Fölsing, *Albert Einstein: A Biography* (New York: Viking, 1997), p. 555.

2. Ibid.

3. Ibid., p. 557.

4. Ibid., p. 563.

5. Ibid., p. 561.

6. Barry Parker, *Search for a Supertheory: From Atoms to Superstrings* (New York: Plenum, 1987), p. 66.

7. Ibid., p. 89.

8. Ibid., p. 65.

9. Lee Smolin, *Three Roads to Quantum Gravity* (New York: Basic Books, 2001), p. 9.

10. Ibid., p. 125.

11. Ibid., p. 115.

12. Ibid., p. 169.

13. Ibid., p. 82.

14. Ibid., p. 138.

15. Ibid., p. 105.

CHAPTER 11. STRINGS AND SUPERSTRINGS

1. Barry Parker, *Search for a Supertheory: From Atoms to Superstrings* (New York: Plenum, 1987), p. 249.

2. Brian Greene, *The Elegant Universe: Superstrings, Hidden Dimensions, and the Quest for the Ultimate Theory* (New York: Norton, 1999), p. 181.

3. Ibid., p. 138.

4. Parker, *Search for a Supertheory*, p. 253.

5. Ibid., p. 255.

6. Ibid.

7. Greene, *Elegant Universe*, p. 139.

8. Ibid., p. 148.

9. Ibid., p. 173.

10. Ibid., p. 222.

11. Ibid., p. 184.

12. Ibid., p. 306.

CHAPTER 12. BEYOND SUPERSTRINGS: M-THEORY

1. Barry Parker, *Search for a Supertheory: From Atoms to Superstrings* (New York: Plenum, 1987), p. 90.

2. Barry Parker, *Einstein's Dream: The Search for a Unified Theory of the Universe* (New York: Plenum, 1986), p. 90.

3. Parker, *Search for a Supertheory*, p. 98.

4. Brian Greene, *The Elegant Universe: Superstrings, Hidden Dimensions, and the Quest for the Ultimate Theory* (New York: Norton, 1999), p. 291.

5. Ibid., p. 140.

6. Ibid., p. 298.

7. Ibid., p. 307.

8. Ibid., p. 312.

9. Ibid.

10. Stephen Hawking, *The Universe in a Nutshell* (New York: Bantam, 2001), p. 54.

11. Michael Duff, "The Theory Formerly Known as Strings," *Scientific American* (February 1998): 64.

12. Greene, *Elegant Universe*, p. 338.

13. Leonard Susskind, "Black Holes and the Information Paradox," *Scientific American,* special ed. (May 2003): 18.

Glossary

ABSOLUTE MOTION. Motion that is the same regardless of the system in which it is measured.

ABSOLUTE TIME. A universal time that is the same for all observers in the universe, independent of their motion.

ABSOLUTE ZERO. The lowest possible temperature.

ACCELERATION. The rate of change of velocity.

ANISOTROPY. Whether the universe is the same in all directions. In this case it is not the same.

ANOMALY. An irregularity.

ANTIMATTER. Matter consisting of antiparticles such as antiprotons.

ASTEROID. A minor object in a solar system. Generally smaller than a moon.

AXIOM. A self-evident truth.

BIG BANG THEORY. A theory of the creation of the universe. Assumes the universe began as an explosion.

BINARY PULSAR. A particular system of two pulsars in orbit around one another.

BLUESHIFT. A shift of spectral lines toward the blue end of the spectrum. It indicates approach.

BOSON. A particle with integral spin.

BRANE. A p-brane has length in p dimensions. A 1-brane is a string, A 2-brane is a sheet. A 3-brane is a volume, and so on.

BROWN DWARF. An object that has slightly less mass than is needed to create a star.

CAUSALITY. The principle that states that cause must come before effect.

CENTER OF MASS. The "average" position of a group of massive bodies. It lies within them.

CEPHEID. A variable star, or one that changes in brightness. Its period is between one and fifty days.

CHAIN REACTION. A reaction in which the fission of one atomic nucleus gives off enough neutrons to cause the fission of another nucleus.

CLASSICAL THEORY. Any nonquantum theory. Relativity theory is an example.

CLUSTER. A group of stars that is held together by their mutual gravitational attraction.

COMPACTIFICATION. A "curling up" of a higher dimension.

COSMOLOGICAL CONSTANT. Constant that Einstein added to the equations of general relativity to stabilize his model of the universe.

COSMOLOGY. A study of the structure of the universe. Usually includes evolution.

COVARIANCE. Concept that implies that the form of equations remains the same under any transformation.

DENSITY. Mass per unit volume.

DOPPLER SHIFT. A change in wavelength that occurs when a body that emits waves is either approaching or receding.

DUALITY. Two ways of looking at something.

ELECTRIC FIELD. The field around a charged particle.

ELECTRODYNAMICS. The study of the interactions of charged particles.

ELECTROMAGNETIC FORCE. The force that arises between two charged particles.

ELECTROMAGNETIC WAVE. A wave given off by an oscillating electric charge.

ELECTRON. The basic particle of electric current. Negatively charged component of the atom.

ENTROPY. A measure of the disorder of a system.

ERGOSPHERE. The region between the event horizon and the static limit of a black hole.

ESCAPE VELOCITY. The velocity needed to overcome a particular gravitational field.

ETHER. A hypothetical substance believed at one time to permeate the universe. Needed to propagate electromagnetic waves.

EUCLIDEAN GEOMETRY. Geometry devised by Euclid, based on axioms and theorems.

EVENT HORIZON. The surface of a black hole. A one-way surface.

EXCHANGE PARTICLE. A particle that is passed back and forth in interactions. The photon is the exchange particle of the electromagnetic interactions.

EXOTIC MATTER. Matter needed to stabilize a wormhole.

FERMION. A particle with a half-integral spin.

FREQUENCY. The number of vibrations per second.

FUSION. The coming together of particles to create energy.

GALAXY. A system consisting of billions of stars.

GEODESIC. The shortest distance between any two points. It can also be the longest.

GLUON. The exchange particle of the strong interactions.

GRAVITATIONAL RADIUS. The radius at which the escape velocity is equal to the velocity of light.

GRAVITON. The hypothetical exchange particle of the gravitational field.

HADRON. A class of particles made up of particles that participate in the strong interactions.

HUBBLE CONSTANT. The constant that gives a measure of how fast galactic velocities increase with distance.

INERTIA. Resistance to change of motion.

INFLATION. A sudden, very dramatic expansion that may have occurred in the early universe.

INTERFERENCE. The interfering of two superimposed waves. They can interfere constructively or destructively.

INTERFEROMETER. An instrument that measures interference.

ISOTROPIC. The same in all directions.

KERR BLACK HOLE. A spinning black hole.

KINETIC ENERGY. The energy of motion.

LASER. An instrument that gives off a coherent beam of light.

LEPTON. A light particle such as an electron.

MAGNETIC MONOPOLE. A heavy particle with either a south or north pole, but not both.

MASS. A measure of the amount of matter in a body.

MATRIX. An array of numbers used in mathematics.

MESON. A medium-heavy particle.

MOMENTUM. Mass multiplied by velocity. A measure of the inertia of an object.

MUON. A particle similar to an electron, but heavier.

NEUTRINO. A massless or very light particle associated with the weak interactions.

NEUTRON STAR. A star made up mostly of neutrons. Very dense and compact.

PERTURBATION. A small disturbance or change.

PHOTOELECTRIC EFFECT. The emission of electrons from a metal when light is shone on it.

PHOTON. A particle of light. The exchange particle of the electromagnetic interactions.

PHOTON SPHERE. The surface around a black hole. Lies 1.5 times farther out than the event horizon.

PLANCK LENGTH. Approximately 10^{-33} cm.

POPULATION INVERSION. When a higher energy level has more particles in it than a lower energy level.

PRECESSION. A slow change of the orientation of the major axis of an elliptical orbit.

PRIMORDIAL BLACK HOLE. A black hole created in the big bang explosion.

PROTON. A basic component of the nucleus of the atom. Positively charged.

PULSAR. A short-period variable star composed of neutrons.

QUANTUM CHROMODYNAMICS. The quantum field theory that describes interactions between quarks and gluons.

QUANTUM FOAM. A foamlike froth that presumably existed in the very early universe, in the time before it was describable by general relativity.

QUANTUM GRAVITY. The union of general relativity and quantum theory.

QUANTUM MECHANICS. The theory of atoms and molecules, their interactions with each other, and their interactions with radiation.

QUASAR. A starlike object with a large redshift that is a strong source of radio waves.

RADIAL VELOCITY. Velocity toward or away from a source.

RADIATION. Electromagnetic energy or photons.

RADIO SOURCE. A source of radio waves. Usually a star or galaxy.

RECESSIONAL VELOCITY. Velocity away from an observer.

REDSHIFT. A shift of spectral lines toward the red end of the spectrum. It indicates recession.

RING SINGULARITY. A singularity in the form of a ring. The Kerr black hole has a ring singularity.

SCHWARZSCHILD RADIUS. The radius at which the escape velocity is equal to the velocity of light.

SINGULARITY. A region where a theory goes awry and gives incorrect answers. Point or ring in the center of a black hole.

SPACE-TIME. A four-dimensional unification of space and time.

SPECTROSCOPE. An instrument for observing spectral lines.

SPECTRUM. The lines seen when light is passed through a spectroscope.

STATIC LIMIT. The region near a black hole where it is impossible to be stationary.

STATISTICS. Numerical facts systematically collected.

STEADY STATE THEORY. The cosmology that assumes that the universe is in a steady state, in other words, always the same.

STRONG NUCLEAR FORCE. The force that holds the nucleons (protons and neutrons) together in the nucleus.

SUPERCOOLING. Cooling below the freezing point without freezing taking place.

SUPERGRAVITY. An early merger of general relativity and quantum theory. It was not generally successful.

SUPERSTRING. Strings within string theory that have supersymmetry.

SUPERSYMMETRY. A principle that relates the properties of particles of different spin.

TACHYON. A hypothetical particle that travels only at speeds greater than that of light. Has never been observed.

TAU. The heaviest lepton.

TENSOR. A component of a branch of mathematics called tensor analysis. The equations of general relativity are written in tensors.

THERMODYNAMICS. The branch of physics that deals with the dynamics of heat.

THERMONUCLEAR. Nuclear reactions that give off heat in the core of a star.

TIDAL FORCE. The force on a body due to a varying gravitational field.

TIME DILATION. A decrease in time interval caused by motion.

TRANSFORMATION. A mathematical relation between two systems. A change of coordinates.

UNCERTAINTY PRINCIPLE. A principle that states that there is an uncertainty when we attempt to measure two or more variables simultaneously in physics.

UNIFIED FIELD THEORY. An attempt to include electromagnetism in general relativity, or more generally, to unify all fields.

VIRTUAL PAIR. A particle-antiparticle pair that appears briefly out of the vacuum, then disappears.

WAVELENGTH. The length between two equivalent positions on a wave.

WHITE DWARF. A small dense star, slightly larger than Earth.

WHITE NEBULA. Early term to describe white, diffuse objects in the sky.

WORLD LINE. The line of "events" or position of an observer over time.

WORMHOLE. A warped space in the form of a tunnel leading up to a black hole.

W PARTICLE. The exchange particle of the weak interactions.

Bibliography

CHAPTER 1. TWISTS IN THE FABRIC OF SPACE

Bernstein, Jeremy. *Einstein.* New York: Viking, 1973.

Brian, Denis. *Einstein: A Life.* New York: Wiley, 1996.

Clark, Ronald. *Einstein: The Life and Times.* New York: World, 1971.

Fölsing, Albrecht. *Albert Einstein: A Biography.* New York: Viking, 1997.

Frank, Philipp. *Einstein: His Life and Times.* New York: Knopf, 1972.

Highfield, Roger, and Paul Carter. *The Private Lives of Albert Einstein.* London: Faber and Faber, 1993.

Hoffmann, Banesh. *Albert Einstein: Creator and Rebel.* New York: Viking, 1972.

Overbye, Dennis. *Einstein in Love.* New York: Penguin, 2001.

Pais, Abraham. *Subtle Is the Lord: The Science and the Life of Albert Einstein.* New York: Oxford University Press, 1982.

Parker, Barry. *Einstein: The Passions of a Scientist.* Amherst, NY: Prometheus Books, 2003.

CHAPTER 2. EXPANDING TO SPACE: HISTORY OF THE COSMOLOGICAL CONSTANT

Aczel, Amir. *God's Equation: Einstein, Relativity, and the Expanding Universe.* New York: Four Walls Eight Windows, 1999.

Ferris, Timothy. *Coming of Age in the Milky Way.* New York: Doubleday, 1988.

Gardner, Martin. *The Relativity Explosion.* New York: Vintage, 1976.

Horgan, Craig, et al. "Special Report: Revolution in Cosmology." *Scientific American* (January 1999).

Milgrom, Mordehai. "Does Dark Matter Really Exist?" *Scientific American* (August 2002): 42.

Parker, Barry. *Creation: The Story of the Origin and Evolution of the Universe.* New York: Plenum, 1988.

———. *The Vindication of the Big Bang: Breakthroughs and Barriers.* New York: Plenum, 1988.

Trefil, James. *The Moment of Creation: Big Bang Physics from Before the First Millisecond to the Present Universe.* New York: Scribners, 1983.

CHAPTER 3. BLACK HOLES, WORMHOLES, AND OTHER DEMONS

Al-Khalili, Jim. *Black Holes, Wormholes, and Time Machines.* Bristol: Institute of Physics, 1999.

Boslough, John. *Stephen Hawking's Universe.* New York: Avon, 1985.

Gribbin, John. *White Holes: Cosmic Gushers in the Universe.* New York: Delacortes, 1977.

Hawking, Stephen. *A Brief History of Time.* New York: Bantam, 1987.

Parker, Barry. *Cosmic Time Travel: A Scientific Odyssey.* New York: Plenum, 1991.

Pickover, Clifford. *Black Holes: A Traveler's Guide.* New York: Wiley, 1996.

Thorne, Kip. *Black Holes and Time Warps: Einstein's Outrageous Legacy.* New York: Norton, 1994.

CHAPTER 4. THE MYSTERY OF TIME AND TIME TRAVEL

Al-Khalili, Jim. *Black Holes, Wormholes, and Time Machines.* Bristol: Institute of Physics, 1999.

Davies, Paul. *How to Build a Time Machine*. New York: Viking, 2001.

Gott, Richard. *Time Travel in Einstein's Universe*. Boston: Houghton Mifflin, 2001.

Macvey, John. *Time Travel*. Chelsea: Scarborough House, 1990.

Morris, Richard. *Time's Arrows: Scientific Attitudes toward Time*. New York: Simon and Schuster, 1984.

Parker, Barry. *Cosmic Time Travel: A Scientific Odyssey*. New York: Plenum, 1991.

Stix, Gary, et al. "Special Issue: A Matter of Time." *Scientific American* (September 2002): 36.

CHAPTER 5. RIPPLES IN THE CURVATURE OF SPACE

Bartusiak, Marcia. *Einstein's Unfinished Symphony: Listening to the Sounds of Space-Time*. Washington, DC: Joseph Henry Press, 2000.

Gibbs, W. Wayt. "Ripples in Spacetime." *Scientific American* (April 2002): 62.

Thorne, Kip. *Black Holes and Time Warps: Einstein's Outrageous Legacy*. New York: Norton, 1994.

CHAPTER 6. GRAVITY'S COSMIC LENSES

Cohen, Nathan. *Gravity's Lens: Views of the New Cosmology*. New York: Wiley, 1988.

Schneider, Peter, Jürgen Ehlers, and Emilio Falco. *Gravitational Lenses*. New York: Springer, 1999.

Wambsganss, Joachim. "Gravity's Kaleidoscope." *Scientific American* (November 2001): 64.

CHAPTER 7. EINSTEIN'S QUANTUM LEGACY

Aczel, Amir. *Entanglement: The Greatest Mystery in Physics*. New York: Four Walls Eight Windows, 2001.

Fine, Arthur. *The Shaky Game: Einstein, Realism, and the Quantum Theory*. Chicago: University of Chicago Press, 1996.

Greene, Brian. *The Elegant Universe: Superstrings, Hidden Dimensions, and the Quest for the Ultimate Theory*. New York: Norton, 1999.

Gribbin, John. *In Search of Schrödinger's Cat: Quantum Mechanics and Reality*. New York: Bantam, 1984.

Pagels, Heinz. *The Cosmic Code: Quantum Physics as the Language of Science.* New York: Simon and Schuster, 1982.

CHAPTER 8. SUPERBOMBS

Bodanis, David. *E = mc²: A Biography of the World's Most Famous Equation.* New York: Berkeley Books, 2000.

Fölsing, Albrecht. *Albert Einstein: A Biography.* New York: Viking, 1997.

Parker, Barry. *Quantum Legacy: The Discovery That Changed Our Universe.* Amherst, NY: Prometheus Books, 2002.

Rhodes, Richard. *Dark Sun: The Making of the Hydrogen Bomb.* New York: Simon and Schuster, 1995.

————. *The Making of the Atomic Bomb.* New York: Simon and Schuster, 1986.

CHAPTER 9. OTHER EINSTEIN INSIGHTS

Cornell, Eric, and Carl Wieman. "The Bose-Einstein Condensate." *Scientific American* (March 1998): 40.

Dannen, Gene. "The Einstein-Szilard Refrigerators." *Scientific American* (January 1997): 90.

Fölsing, Albrecht. *Albert Einstein: A Biography.* New York: Viking, 1997.

Griffin, A., D. Snoke, and S. Stringari, eds. *Bose-Einstein Condensation.* Cambridge: Cambridge University Press, 1995.

CHAPTER 10. DREAMS OF A UNIFIED THEORY

Brian, Denis. *Einstein: A Life.* New York: Wiley, 1996.

Fölsing, Albrecht. *Albert Einstein: A Biography.* New York: Viking, 1997.

Greene, Brian. *The Elegant Universe: Superstrings, Hidden Dimensions, and the Quest for the Ultimate Theory.* New York: Norton, 1999.

Parker, Barry. *Einstein: The Passions of a Scientist.* Amherst, NY: Prometheus Books, 2003.

Smolin, Lee. *Three Roads to Quantum Gravity.* New York: Basic Books, 2001.

CHAPTER 11. STRINGS AND SUPERSTRINGS

Duff, Michael. "The Theory Formerly Known as Strings." *Scientific American* (February 1998): 64.

Greene, Brian. *The Elegant Universe: Superstrings, Hidden Dimensions, and the Quest for the Ultimate Theory.* New York: Norton, 1999.

Hawking, Stephen. *The Universe in a Nutshell.* New York: Bantam, 2001.

CHAPTER 12. BEYOND SUPERSTRINGS: M-THEORY

Duff, Michael. "The Theory Formerly Known as Strings." *Scientific American* (February 1998): 64.

Greene, Brian. *The Elegant Universe: Superstrings, Hidden Dimensions, and the Quest for the Ultimate Theory.* New York: Norton, 1999.

Hawking, Stephen. *The Universe in a Nutshell.* New York: Bantam, 2001.

Index